CW01163529

Papermaking Science and Technology

a series of 19 books
covering the latest
technology and
future trends

Book 19

Environmental Control

Series editors
Johan Gullichsen, Helsinki University of Technology
Hannu Paulapuro, Helsinki University of Technology

Book editor
Pertti Hynninen, Enviro Data Oy

Series reviewer
Brian Attwood, St. Anne's Paper and Paperboard Developments Ltd

Book reviewers
Kathleen M. Bennett, Fort James Corporation
Allan M. Springer, Miami University

Published in cooperation with the Finnish Paper Engineers' Association and TAPPI

ISBN 952-5216-00-4 (the series)
ISBN 952-5216-19-5 (book 19)

Published by Fapet Oy
(Fapet Oy, PO BOX 146, FIN-00171 HELSINKI, FINLAND)

Copyright © 1998 by Fapet Oy. All rights reserved.

Gummerus Printing, Jyväskylä, Finland 1998

Printed on LumiMatt 100 g/m2, Enso Fine Papers Oy, Imatra Mills

Foreword

Johan Gullichsen and Hannu Paulapuro

PAPERMAKING SCIENCE AND TECHNOLOGY

Papermaking is a vast, multidisciplinary technology that has expanded tremendously in recent years. Significant advances have been made in all areas of papermaking, including raw materials, production technology, process control and end products. The complexity of the processes, the scale of operation and production speeds leave little room for error or malfunction. Modern papermaking would not be possible without a proper command of a great variety of technologies, in particular advanced process control and diagnostic methods. Not only has the technology progressed and new technology emerged, but our understanding of the fundamentals of unit processes, raw materials and product properties has also deepened considerably. The variations in the industry's heterogeneous raw materials, and the sophistication of pulping and papermaking processes require a profound understanding of the mechanisms involved. Paper and board products are complex in structure and contain many different components. The requirements placed on the way these products perform are wide, varied and often conflicting. Those involved in product development will continue to need a profound understanding of the chemistry and physics of both raw materials and product structures.

Paper has played a vital role in the cultural development of mankind. It still has a key role in communication and is needed in many other areas of our society. There is no doubt that it will continue to have an important place in the future. Paper must, however, maintain its competitiveness through continuous product development in order to meet the ever-increasing demands on its performance. It must also be produced economically by environment-friendly processes with the minimum use of resources. To meet these challenges, everyone working in this field must seek solutions by applying the basic sciences of engineering and economics in an integrated, multidisciplinary way.

The Finnish Paper Engineers' Association has previously published textbooks and handbooks on pulping and papermaking. The last edition appeared in the early 80's. There is now a clear need for a new series of books. It was felt that the new series should provide more comprehensive coverage of all aspects of papermaking science and technology. Also, that it should meet the need for an academic-level textbook and at the same time serve as a handbook for production and management people working in this field. The result is this series of 19 volumes, which is also available as a CD-ROM.

When the decision was made to publish the series in English, it was natural to seek the assistance of an international organization in this field. TAPPI was the obvious partner as it is very active in publishing books and other educational material on pulping and papermaking. TAPPI immediately understood the significance of the suggested new series, and readily agreed to assist. As most of the contributors to the series are Finnish, TAPPI provided North American reviewers for each volume in the series. Mr. Brian Attwood was appointed overall reviewer for the series as a whole. His input is gratefully acknowledged. We thank TAPPI and its representatives for their valuable contribution throughout the project. Thanks are also due to all TAPPI-appointed reviewers, whose work has been invaluable in finalizing the text and in maintaining a high standard throughout the series.

A project like this could never have succeeded without contributors of the very highest standard. Their motivation, enthusiasm and the ability to produce the necessary material in a reasonable time has made our work both easy and enjoyable. We have also learnt a lot in our "own field" by reading the excellent manuscripts for these books.

We also wish to thank FAPET (Finnish American Paper Engineers' Textbook), which is handling the entire project. We are especially obliged to Ms. Mari Barck, the project coordinator. Her devotion, patience and hard work have been instrumental in getting the project completed on schedule.

Finally, we wish to thank the following companies for their financial support:

A. Ahlstrom Corporation
Enso Oyj
Kemira Oy
Metsä-Serla Corporation
Rauma Corporation
Raisio Chemicals Ltd
Tamfelt Corporation
UPM-Kymmene Corporation

We are confident that this series of books will find its way into the hands of numerous students, paper engineers, production and mill managers and even professors. For those who prefer the use of electronic media, the CD-ROM form will provide all that is contained in the printed version. We anticipate they will soon make paper copies of most of the material.

List of Contributors

Pertti Hynninen, Dr. Tech, Docent

Enviro Data Oy

Helsinki University of Technology

Preface
Pertti Hynninen

In today's forest industry, environmental protection is part of everyone's duties. The work of environmental management, as it is frequently referred to, is taking up more and more time, not least in solving related technical and economic issues. However, this work is now being done to increasing effect. The purpose of this volume is to provide a broadly based source of information for those working in this field. Given the rate at which environmental protection is moving, readers and others are constantly in need of new information. This need has been addressed in this book.

The compilation of this book has involved numerous people. The contribution of Pertti Laine, Director at the Finnish Forest Industries Federation, has been of major importance in relation to the content as well as to the material included in the appendices.

I wish to thank all those who took part in this work.

Pertti Hynninen

Table of Contents

1. Introduction 10
2. Environmental controls 13
3. Environmental permits for industry 27
4. Effluent loadings from the forest industry 31
5. Raw water treatment 43
6. Effluent treatment 57
7. Reducing emissions to air 95
8. Solid and liquid wastes 109
9. Other environmental impacts and their reduction 133
10. Tools for environmental management 137
11. Appendix 1. 143
12. Appendix 2. 208
13. Appendix 3. 216
14. Appendix 4. 221
15. Appendix 5. 226
    Conversion factors 228

CHAPTER 1

Introduction

CHAPTER 1

Introduction

Effluent loadings and sulfur emissions from the forest industry came under considerable scrutiny in Finland in the 1970s. Actions taken to further develop the processes and to deal with emissions brought the results shown in Figs. 1 and 2[1].

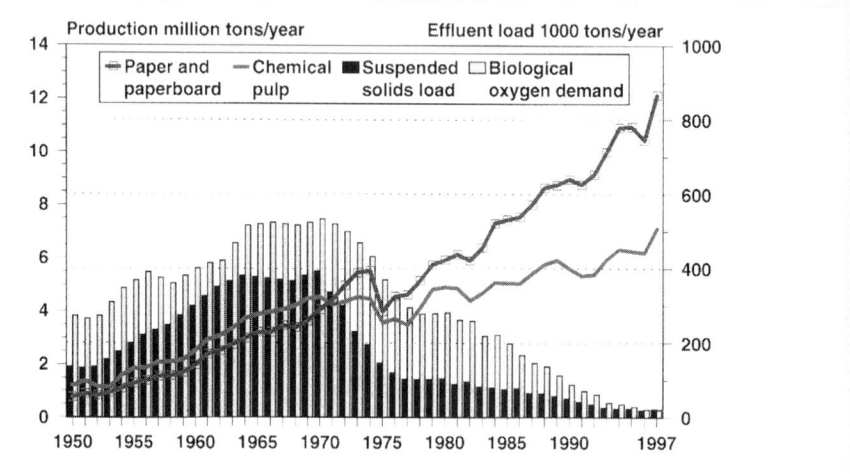

Figure 1. Trend in effluent loadings from the Finnish forest industry.

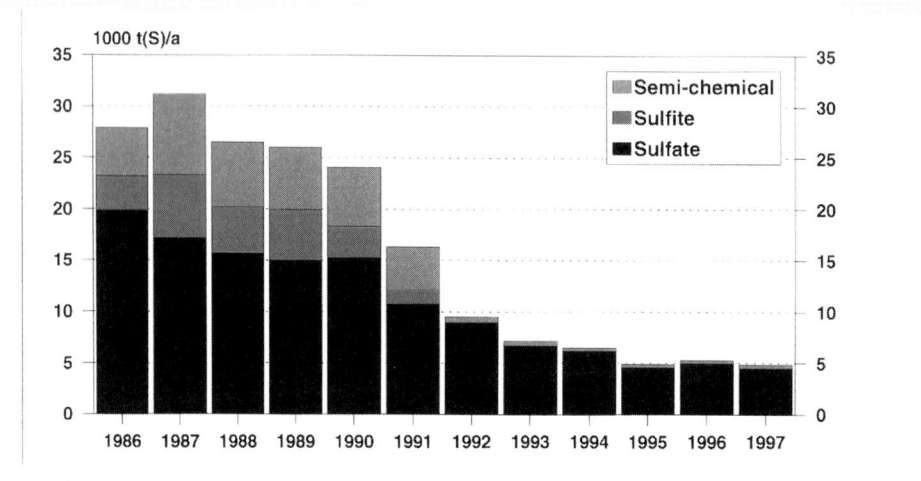

Figure 2. Trend in sulfur emissions from Finnish pulp mills.

Introduction

The fall in emissions corresponds with other established forest industry countries in Europe, North America, and Asia. The reasons for the decrease in emissions are changes in production technology and in the products themselves. For example, the industry has discontinued the production of calcium sulfite pulp, which causes high emissions. The energy crises of the 1970s and early 80s forced the introduction of numerous measures that resulted in reductions in mill water consumption and emissions to the environment. Legislation and regulatory controls also have tightened, particularly in relation to effluents. As forest industry units are major point sources of effluent loadings, it is often a simple matter to determine the effects of reductions in their loadings. The increasing importance attached to the quality of the environment in recent years also has prompted nongovernmental organizations to press for lower emissions and for multiple use of forests.

Despite this, the debate over environmental pollution has lost none of its intensity. Demands for an end to the use of chlorine, to the clearcutting of forests, and to the pollution of soil by wood preservatives, in addition to other somewhat sensationalist headlines, have helped shape public opinion. The reasons for this are found both within the industry and outside it. Much still remains to be done in the area of environmental protection.

Over the past decade, environmental protection in the forest industry has become increasingly integrated into general managementof the mill. Employees have received training, responsibility, and the authority to act in accordance with environmental quality systems.

In this book, the aim is to consider environmental management by reviewing the relevant legislation, ISO 14000 systems, life cycle analysis, and environmental labels. This section on the basics of environmental protection concerns mainly the forest industry. By broadening our understanding of the issues dealt with here, we shall be able to deal with the environmental requirements the industry will have to face in the future. This applies particularly to the production of pulp and paper. Mechanical wood processing and printing, with the special issues these raise, are not discussed.
A short list at the end of Chapter 6 provides recommended reading for those interested in finding out more. The same sources were used in producing the text for this book.

CHAPTER 2

Environmental controls

1	**Regulatory controls**	**14**
1.1	Handling of environmental issues by the EU	14
	1.1.1 Environmental protection policy	15
	1.1.2 Basic principles	15
	1.1.3 Environmental action programs	17
	1.1.4 The IPPC Directive	18
	1.1.5 Eco-Management and Audit Scheme (EMAS)	18
	1.1.6 Environmental labels	18
	1.1.7 Packaging and Waste Packaging Directive	19
	1.1.8 Environment-related information and the European Environment Agency (EEA)	19
1.2	Environmental protection legislation in Finland	19
	1.2.1 Environmental protection administration in Finland	21
2	**Economic instruments**	**22**
3	**Market instruments**	**23**
4	**Other measures**	**23**
	References	24

CHAPTER 2

Environmental controls

Environmental protection is a much wider concept than, say, nature conservation. It means different things to different people, and is constantly changing. The driving force behind environmental protection has always been the desire to secure man's well-being. The methods currently used in directing environmental protection are shown in Fig. 1.

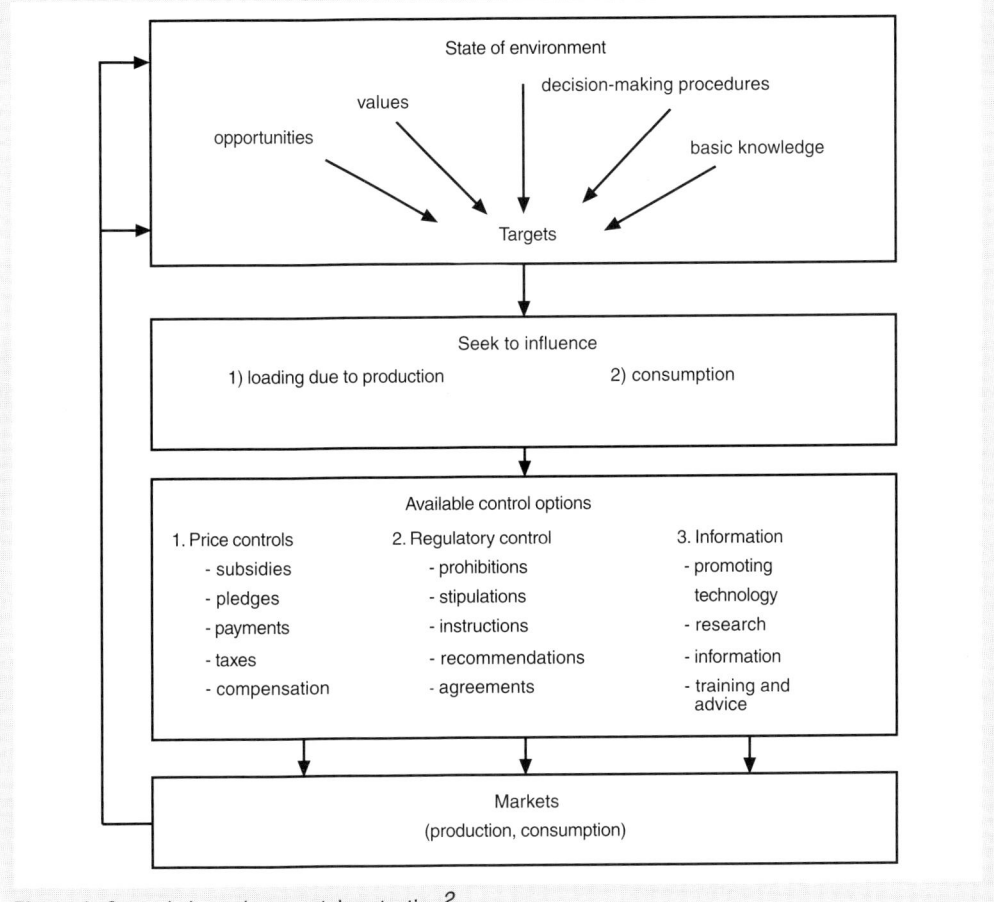

Figure 1. Controls in environmental protection[2].

CHAPTER 2

Originally, the most stringent measures were based on legislation and on enforcement of the resulting statutes and decrees, which in turn were based on stipulations and instructions issued by the authorities. In some respects, this approach is now very much out of date. For example, in certain parts of southern Europe the laws governing the management of irrigation systems contain elements dating back to Roman times. In Finland, the first water legislation concerned the harnessing of rapids for hydropower and lake drainage to produce more agricultural land.

As well as issuing prohibitions and instructions, the government also employs economic controls. In Finland recently, this mainly concerned the debate over the introduction of an environmental tax. The government also provides information, advice, and training when necessary.

In the case of many products, among them paper and oil, market forces have had a major impact on environmental protection. The forest industry has been a fairly easy target in those countries that are the biggest importers of forest products.

1 Regulatory controls

Regulatory controls have been based on legislation, including the grounds for implementing environmental protection in practice. In terms of water pollution control, the legislation is in some cases extremely old. Most of the legislation relating to air pollution control, however, is comparatively recent, dating back only 100–150 years. The following is a review of European Union (EU) environmental legislation, including that affecting the Finnish forest industry, and the thinking behind it. Appendix 1 lists some of the legislative measures introduced in certain prominent forest industry countries like Canada and USA.

1.1 Handling of environmental issues by the EU

At present, the European Union comprises 15 member states. The EU's principal institutions are the Parliament, whose members are elected; the Commission, comprising 20 Commissioners; the Council of Ministers representing each of the member states; and the European Council, which is made up of the heads of state or government of the member states.

The Commission is the EU's executive, as it draws up proposals for legislation and action at the European level and monitors their application in member countries. Its members are approved by the European Parliament. A committee representing 24 Directorates General (DG), each with its own portfolio, assists the Commission in its work. Within the Commission, environmental matters are the responsibility of DG XI. The directorates general for trade (I), industry (III), competition (IV), agriculture (VI), research (XII), internal markets (XV), and energy (XVII) also deal with issues related to the forest industry.

What follows is a general review of EU environmental protection policy (see also Appendix 1). Environmental protection is currently undergoing considerable change and development. The account given in this book reflects the situation in the first half of 1997.

1.1.1 Environmental protection policy

The founding document of the European Union is considered to be the 1957 Treaty of Rome, which established the then European Economic Community. The Treaty set out the principles and aims of the Community, but contained no regulations concerning environmental protection. The first actions of an environmental nature taken by the EEC concerned products and their safety. In 1972, the EEC heads of state put forward an initiative for the formulation of a common environmental policy, and subsequently produced the first environmental action program for the years 1973–77. Environmental policy acquired its present status when the Community's competence and objectives in respect of environmental issues were entered in the Single European Act of 1987. The Community defined its objectives in the sphere of environmental protection as preserving and protecting the environment, improving the quality of the environment, promoting the safeguarding of human health, and the prudent and rational use of natural resources.

The Maastricht Treaty that came into effect in November 1993 stated that one of the EU's tasks is to promote sustainable growth with due regard for the environment.

The EU lists environmental protection as one of the its basic policy areas alongside those relating to trade policy, competition, etc. The Treaty also requires environmental issues to be taken into account in all the EU's different spheres of activities in both planning and implementation. The Treaty provides for environmental issues to be decided, in most cases, by a majority vote, and this has speeded up the relevant decision-making.

1.1.2 Basic principles

Article 130r of the Treaty of Rome sets out the objectives of the EU's environmental policy by stating that Community policy on the environment shall contribute to pursuit of the following objectives:

- Preserving, protecting, and improving the quality of the environment
- Protecting human health
- Prudent and rational utilization of natural resources
- Promoting measures at international level to deal with regional or worldwide environmental problems.

Certain principles guide the EU's environmental policy by serving as boundary conditions and guidelines for protection measures. Article 130r of the Treaty of Rome also states these principles. Perhaps the most significant addition to these is the principle of sustainable development, as stated in the Treaty of Amsterdam.

High level of protection

Policy on the environment must aim at a high level of protection. The Commission must base its proposals for harmonization of environmental protection on a high level of protection. In applying this principle, account must be taken of the diversity of situations in the different regions. This relativity specifically concerns the definition of a high level of protection, as what is regarded as high in this context can vary from one region to another. This allows different protection standards to be applied in different regions. The principle is rather general and its main purpose can be taken to be the desire to elevate the status of environmental protection.

Principle of prevention

The principle of prevention refers to environmental hazards whose likelihood of occurrence is known with some degree of probability. This means that requesting certain protective measures in advance can prevent hazards from arising at all. The design of this principle is to help protect the environment at as early a stage as possible and to encourage prevention rather than reacting to hazards after they have occurred.

Precautionary principle

The precautionary principle requires precautions to be taken to prevent hazards arising from human activity. It goes further than the principle of preventive action in that it requires that action be taken before it can be assumed that some accident or consequence might arise. The principle does not require proof that an environmental hazard exists; instead it allows the EU to take action before there is any definite proof of the cause-effect relationship between human activity and damage.

Source principle

If an activity causes damage to the environment, the damage must be tackled as quickly as possible, i.e., at source. Application of this principle is when the activity in question results in such substantial benefits that it cannot be discontinued despite the harm caused to the environment. Its use also can try to prevent damage to the environment across national frontiers. Interpretation of the principle at a more general level based thinking on the idea that it is more expedient to tackle environmental hazards through controls aimed at the polluter than through actions directed at the affected area. Combating pollution at source also applies a geographical basis, for example, in waste management legislation, which states that wastes should be dealt with as close to their source as possible.

Polluter pays principle

This principle states that the party whose activity causes the hazard should bear the cost of reducing and preventing environmental hazards. It is based on the thinking that no one has the right to pollute the environment at others' expense. The principle also intends that the cost of preventing pollution should be passed on to product prices, thereby steering consumption toward products that cause smaller environmental loadings.

Environmental controls

In the European Union, environmental protection also should take into account the following matters stated in Article 130r:

- Available scientific and technical data
- Environmental conditions in the various regions of the Community
- The potential benefits and costs of action or lack of action
- The economic and social development of the Community as a whole and the balanced development of its regions.

The intention is that the EU should be aware of the scope and consequences of its work to protect the environment. These matters must be taken into account in the planning both of environmental programs and of individual measures.

The Maastricht Treaty requires environmental considerations to be taken into account in all areas of activity. This means that decision-making in other areas (such as industrial and agricultural policy) should consider the environmental implications of the plans made and actions taken. This emphasized status of environmental protection is unique within the EU: for no other objective is there subject a comprehensive obligation for fulfilment.

In the issuing of regulations, all stages – from planning to final implementation – must apply the integration principle. However, environmental protection does not automatically take precedence over other objectives; in some cases, even activities that are detrimental to the environment can be tolerated. The essential point of this principle is that the consideration of each case takes into account environmental protection requirements.

1.1.3 Environmental action programs

In seeking to protect the environment, the European Union has introduced a number of environmental action programs, each spanning several years. The fifth such environmental action program "Towards Sustainability" received approval in 1993. It differs from previous programs in that it takes a far more global approach in line with the EU's increasing role in environmental protection both regionally and globally. At the same time, the program sets longerterm objectives.

In the Fifth Environmental Action Program the emphasis is on preventing problems rather than on rectifying them once they have arisen. It includes the objective of steering consumption and consumer behavior in a more environment-friendly direction. It seeks to achieve greater sustainability not just at the Community level, but also in member states, in trade and industry, and among ordinary people by encouraging everyone to take a more active role in protecting the environment. An interim report on the Fifth Environmental Action Program notes that progress has been made in integrating environmental considerations into the different target sectors – industry, energy, transport, agriculture, and tourism – but at different rates. Progress is most advanced in manufacturing industry but least apparent in agriculture and tourism.

1.1.4 The IPPC Directive

The Integrated Pollution Prevention and Control (IPPC) Directive states that the permits issued to manufacturing plants should take account of the full spectrum of emissions attributable to manufacture. The IPPC Directive requires that activities which harm the environment be subject to a permit. The conditions for the granting of such a permit relate to both air and water pollution control, waste disposal, and noise abatement. The Directive intends not merely to reduce emissions throughout the entire area of the EU but also to promote the use of environment-friendly technology. The IPPC Directive states that the permit conditions applied to industrial manufacturing plants must be based on the use of Best Available Techniques (BAT), but account must also be taken of the plant's technical features, economics, geographical location, and local environmental circumstances. In the forest industry, what constitutes BAT will obviously have to be decided as accurately as possible for each industrial plant individually.

Implementation of the IPPC Directive has meant changing environmental legislation and permit arrangements in member states. These changes must all be in force by 1999.

1.1.5 Eco-Management and Audit Scheme (EMAS)

The design of the EU's Eco-Management and Audit Scheme (EMAS) is to encourage companies to introduce environmental management systems on a voluntary basis. EMAS is part of attempts by the EU to shift the environmental pressure on companies away from statutory controls and to make it more subject to market forces. The environmental policy, objectives, program, review, and audit are all internal matters for the company concerned. However, for the company to gain EMAS registration, an external accredited verifier must review and validate these internal matters. The company must publish an environmental policy statement, and this too requires approval by an external verifier. EMAS means that environmental protection becomes an integral part of a company's overall operations (see also the ISO 14000 environmental management system, Appendix 2).

1.1.6 Environmental labels

The purpose of environmental labels is to provide consumers with more information about the environmental impacts of products and to steer product manufacture and consumption toward more environment-friendly solutions. Environmental labels are awarded only to those products that can be shown to impose smaller loads on the environment than other products intended for the same purpose. Most environmental label systems have so far been national, one notable exception being the Nordic Environmental Swan Label. Criteria for awarding the Swan label have now been approved for about 50 product groups, more than ten of which are forest industry products. The EU has been developing its own environmental labeling system since 1992, and criteria for 11 product groups had been approved by November 1996. The forest industry products covered are copier paper, household paper, and toilet tissue. (For more on environmental labels, see Appendix 4.)

1.1.7 Packaging and Waste Packaging Directive

The Packaging and Waste Packaging Directive approved by the EU in 1994 has important implications for the forest industry. It aims at harmonizing member states' legislation on packaging and waste packaging, reducing adverse environmental impacts, and securing the free flow of goods within the EU's internal markets. The Directive requires member states to recover at least 50% and at most 65% (by weight) of all waste packaging and to recycle at least 25% and at most 45% of all waste packaging. The remaining waste packaging can be used for purposes such as energy production. These objectives are to be achieved by summer 2001.

1.1.8 Environment-related information and the European Environment Agency (EEA)

The free availability of information on the environment is the subject of a Directive issued in 1990. Its aim is to ensure that everyone has the opportunity to obtain information freely on the state of the environment, on environmental protection, and on activities that have a harmful effect on the environment. The Directive places an obligation on the authorities actively to supply information on the state of the environment. The European Environment Agency (EEA) was set up in Copenhagen in 1994. Its main function is to collect information and produce statistics on the state of the environment in Europe and to evaluate and publish such information. The EEA is currently setting up environmental information centers in different fields within the EU member countries.

1.2 Environmental protection legislation in Finland

The Water Act (264/61) of 1961 is the basis for present water pollution control legislation. The Act sets out conditions and requirements for the use of water resources (water intake stations, fisheries, boat traffic, hydropower, recreational use, discharge areas for effluents).

Section 1, §19 of the Water Act of 1961 states that no actions may be undertaken that will cause pollution in a receiving watercourse without a permit granted by a water rights court. The water rights court may grant such a permit if the petitioner meets the conditions stated in Section 10, §24 of the Act relating to a comparison of interests and to what is considered reasonable in the situation (see Appendix 1). Pollution of groundwater, on the other hand, is in most cases prohibited (Section 1, §22), although amendments to the Act have eased this restriction somewhat.

A statute (283/62, amended 499/80) relating to measures for preventing water pollution also regulates pollution of watercourses and groundwaters. The statute lists those plants which must make notifications to the water authorities.

The 1961 Water Act contains two other important restrictions on the use of water resources. These concern closure (Section 1 §12) and changes (Section 1 §15). Activities that could contravene these restrictions include waterway construction work and water abstraction. The water rights court may grant a permit for construction work provided the petitioner meets conditions regarding a comparison of interests (Section 2 §6). The same applies in the case of water abstraction. Certain amendments of a tech-

CHAPTER 2

nical nature were made to the Act in the late 1980s, and a more fundamental revision can be expected largely as a result of EU legislation.

The Air Pollution Control Act of 1982 (67/82) is the basis of air pollution control. The related statute (716/82) establishes stipulations regarding notification procedures for air pollution control at industrial plants, together with the monitoring and inspection of emissions to air. The Council of State decision 160/87 provides general guidelines for emissions of gaseous sulfur compounds from sulfate pulp mills. The 1996 amendment to the Act removed the notification procedure by incorporating permit applications into the new environmental permit procedure.

The Waste Management Act (673/78) dates from 1978. The related statute (307/79) stipulates the conditions and requirements for the collection and handling of difficult wastes, establishment of landfill sites, prevention of soil pollution, and the collection and transport of solid wastes. The regulations contained in the new Waste Management Act of 1994 are more detailed. While retaining the previous requirements, the new act prescribes preventive measures – for example, for preventing and reducing the formation of waste and for rendering waste less dangerous and hazardous. The new regulation required industrial plants to reformulate their waste management plans accordingly by the end of 1996.

The Noise Abatement Act (382/87) of 1987 makes those parties causing noise responsible for reducing noise levels the extent considered reasonable, and at the same time to be sufficiently aware of the noise caused by their activities. National Board of Health circular no. 1676 presents recommendations for the highest noise levels permissible in different situations. These recommendations usually form a basis for the conditions set during siting procedures. Under the Noise Abatement Act, the Council of State has the power to issue general instructions and stipulations about noise levels, noise emissions from machinery and transport vehicles, noise abatement measures, and about prohibiting or restricting noisy activity or the use of noisy machinery at specified times. The Council of State also has the power to issue general instructions concerning the zones required to achieve noise abatement.

The Act on Environmental Permit Procedures, which came into effect on September 1, 1992, signified a departure from earlier environmental legislation. It integrated the processing and supervision of siting decisions under the Adjoining Properties Act, of siting permits under the Public Health Act, of air pollution control notifications under the Air Pollution Control Act, of real estate waste disposal plans under the Waste Disposal Act, and of permits for the handling of difficult wastes. This means that a single authority – either a municipal environmental authority or a regional environment agency – now handles applications for environmental permits. This avoids the multiple processing of applications and facilitates a more consistent assessment of environmental impacts. It has also greatly reduced the number of permits and decisions that used to be required.

Listed below are the permits required in Finland under the previous legislation, in this example for a chemical pulp mill project:

Permit (legislation)	Granting authority
Air pollution control notification (Air Pollution Control Act)	Provincial government
Notification of advance water pollution control action (Statute on advance action)	Water and environment district
Effluent discharge permit (Water Act)	Water rights court
Building permits (Building Act)	Ministry of the Environment Provincial Government Municipal council Building board
Siting permit (Public Health Act)	Municipal health board
Waste disposal plan (Waste Management Act)	Provincial government
Decision by building board (Adjoining Properties Act)	Building board

1.2.1 Environmental protection administration in Finland

The Act on the Number of Ministries within the Council of State and their General Competence (1/83) sets out the duties of the Ministry of the Environment as follows:

"The Ministry of the Environment shall deal with matters concerning environmental protection and nature conservation, the overall planning of water usage, management and water pollution control, as well as other matters related to the administration of waters and the environment and other recreational uses of the environment."

The Ministry of the Environment comprises the following departments:

- General Management Department
- Environmental Protection Department
- Environmental Policy Department
- Land Use Department
- Housing Department.

CHAPTER 2

The Environmental Protection Department largely handles environmental protection as it relates to industry.

The Finnish Environment Institute, a consultative body subordinate to the Ministry of the Environment, works to promote water use and management and water pollution control. At the local level, the authorities are regional environment agencies, which correspond to the earlier system of water and environment districts. The environment agencies oversee the use of waterways, including groundwater, monitor the state of waters and any activities affecting it, and carry out the reviews and final inspections referred to in the Water Act.

They are also responsible for the granting and supervision of permits under the Air Pollution Control Act and the Waste Management Act, a function previously carried out by the provincial governments.

2 Economic instruments

The main economic instruments available to the authorities are compensation procedures for defraying the costs of environmental protection and measures related to taxation.

In seeking to promote environmental protection on a wide front, the authorities have offered the incentive of interest concessions on loans as well as other special loan conditions. In industry, changes to these concessions have lifted a turnover tax on environmental protection investments.

The partial funding of environment-related research and development work also has been a significant form of support.

Other economic instruments have been used depending on the business cycle and other factors.

The recently introduced environmental tax is likely to become the most important economic constraint in the future. This tax is decided by Parliament in conjunction with other taxation.

In some European countries, among them France, emission-related payments have been collected for each waterway region, and the proceeds used to promote water pollution control in that region.

Taxation affects industry in different ways, depending on whether the companies concerned serve mainly the domestic or the export market. For industry serving its domestic market, it is usually easier to pass the extra costs on in full to the consumer, one example being energy production. The export industry may not necessarily be able to adjust its prices.

Before meaningful calculations can be made of the costs of environmental protection as a basis for action, cost information is needed. The OECD has drawn up instructions for such calculations, and several countries also have their own instructions for producing statistics. In some cases, these instructions are comparable on an international basis. Statistics Finland, the official statistics office, produced a calculation system of its own in the beginning of 1990s. Other methods of calculation have also developed from time to time but, before employing these, it is expedient to check the purpose for which they were devised.

Economic instruments are quite clearly effective. In order to ensure that such instruments remain appropriate for their intended purpose it is important that they are properly monitored and revised sufficiently often to take into account factors such as advances in technology.

3 Market instruments

"Market instruments" is used here as a general term to refer largely to the different calls for environmental protection transmitted through the markets. Some examples:

- Legislation introduced on export markets, e.g., demands for the use of recycled fiber
- Environmental labels
- Environmental demands from end-users, examples being calls for an end to the chlorine bleaching of pulp and boycotts on products derived from wood obtained by clear felling or from primary forests.

For industrial manufacturers relying on exports, the situation is often awkward because of the difficulty of establishing a meaningful dialogue and because of the lack of price flexibility. Environmental protection often gets confused with unrelated issues such as the competition between industry in different countries. Nevertheless, the only sustainable approach is to adapt to the situation and allow sufficient time for change to take place.

Market pressure was partly responsible for the introduction of product life cycle analysis, initially in the packaging industry. Life cycle analysis is useful in the planning of products that cause smaller environmental emissions.

In some cases, market forces have been strong enough to outweigh certain guidelines such as those on emissions. The forest industry can probably expect more, and to varying degrees unpredictable, demands from its main markets.

4 Other measures

International agreements often have formed the basis for actions designed to protect the environment in different countries. Of particular importance to Finland is the Baltic Sea Treaty. Wider-ranging agreements in recent years include the Earth Summit in Rio, notably in relation to air pollution control. Such agreements not only influence national legislation but also make it easier to arrange funding for environment-related projects. One way is to ease the burden of borrowing by granting interest rate concessions.

In export trade, measures to prevent subsidies that distort the competitive situation are also important. General recommendations of this kind have been put forward within the OECD.

References

1. Metsäteollisuus Ry., *Effluent loading statistics, 1997.*
2. *Ympäristönsuojelun taloudellinen ohjaus, Komitean mietintö 1989:18, Helsinki, 1989, pp. 17.*

CHAPTER 3

Environmental permits for industry

1 Discharge of effluents to receiving waters 27
2 Location-related environmental permits 28
3 Environmental Impact Assessment (EIA) 29

CHAPTER 3

Environmental permits for industry

1 Discharge of effluents to receiving waters

The practice with regard to permits depends on national or local environmental acts and decrees. Decisions are made as a result of court proceedings or they may take the form of rulings by the relevant authorities. Appeals against such rulings usually lead to court hearings. The following text, with the aid of a few examples, describes the current practice in Finland.

Unlike in other countries, most waters in Finland and Sweden are privately owned. A system of water rights courts has been set up to deal with disputes; the various authorities have no power to issue rulings that would adversely affect the protection of property (see also Appendix 1).

Section 3, §71 of the Water Decree specifies the information that should be included in an application for permission to discharge effluent. The main points are as follows:

Technical information on the process
- Description of process used for product manufacture
- Estimated effluent loading
- An account of measures for reducing the loading

Description of the receiving water
- State of the water and expected changes in it
- The suitability of the water for various uses and how this will be affected

Cost of reducing discharges

Names of those who own the water (as far as applicant is aware)

Ways known to the applicant of completely preventing the expected adverse effects

Applications are submitted to the water rights court in whose area the water in question is located.

In dealing with an application, the water rights court may adopt the procedure of giving public notice of the application, usually in cases where loadings have been reduced. The application is publicly announced and the court bases any further action on complaints or other feedback. While the discharge permit may be granted, measures relating to the payment of compensation are not dealt with in the same way. What usually happens is that the water rights court appoints a suitably qualified engineer to carry out the necessary further investigations. The parties involved (applicant, water owners, and the National Board of Waters and the Environment representing the public interest) normally voice their concerns as part of this procedure. Based on the information it has received, the water rights court issues the technical conditions under which the permit is to be granted and determines compensation to be paid for any damage caused to the environment. In most cases, the technical conditions require a plan for a treatment plant to be submitted to the relevant authorities, discharge limits in kg/d, expressed in terms of the usual effluent parameters (suspended solids, BOD, COD, AOX, and nutrients), and a plan, to be approved by the relevant authority, for monitoring the receiving water. The relevant authority in most cases is the district organization of the National Board of Waters and the Environment. If the parties concerned are not satisfied with the decision, they normally have the right to appeal to Water Rights Appeal Court.

Despite recent improvements, the processing of applications still takes a considerable length of time, especially in regard to investigating and deciding compensation for damage. Today, decisions are reached fairly quickly, taking between six months and two years. This is thanks to changes in the nature of the application. Discharges are diminishing and less time is spent dealing with compensation-related matters. Permits granted by a water rights court are normally valid indefinitely. However, there is usually a clause stating the date by which an application should be made to renew the permit.

2 Location-related environmental permits

Finland has a broad body of legislation concerning the use and protection of the environment, much of it found in different laws passed at different times.

The building or expansion of an industrial plant has always required a large number of permits or approval from the authorities. The following have always been subject either to permits or to legislation in general:

- Choice of construction site and the size and type of building
- Emissions to the air
- Waste management
- Noise abatement
- Public health and safety at work
- Storage and transport of toxic or otherwise dangerous and inflammable substances
- Transport of dangerous substances
- Construction and use of pressure vessels
- Electricity supplies.

Many of the above are also significant for environmental protection. The granting of permits for siting or constructing a building used to require a written account of some or all of the above points. Applications were usually processed at the municipal level, with requests for the necessary statements and rulings. Numerous laws and decrees as well as various recommendations were the basis for this type of control. The practice is still basically the same today, as amended by the requirements of the Environmental Permit Procedures Act (735/91).

Most of the present regulations relating to air pollution control and waste management for existing plants are based on notifications regarding air pollution control and waste management submitted by the plants to the environmental protection departments of provincial governments, and on the resulting rulings.

An air pollution control notification is a document detailing the quantities of emissions and their sources, and what measures are to be taken regarding the emissions. An application must also include an estimate of the cost of lowering emissions. A waste management notification must contain an account of the quantities of waste and explain how the waste can be rendered harmless.

The provincial government has made rulings either approving or amending the proposed measures. The environmental protection departments of provincial governments have now been merged into the regional environment agencies.

With the exception of the discharge of effluents to receiving waters, the system of controls today is based almost entirely on stipulations issued in response to air pollution control and waste management notifications and on the statutory monitoring requirements for environmental protection and environmental impacts set out in siting permits. The environmental permits mentioned earlier are now being introduced as the basis for controls.

3 Environmental Impact Assessment (EIA)

Environmental Impact Assessment (EIA) is a major consideration in arriving at rulings on environmental protection. Applicants normally commission the necessary investigations, the scope of which is usually specified in detail, from companies with the required competence.

In Finland, the act on environmental impact assessment has modified the procedure for EIA.

The aim here has been to bring Finland's EIA procedure into line with procedures adopted in other European countries. The practice concerns pulp and paper mills with a daily production in excess of 200 metric tons.

In terms of their content, present-day application documents differ little from their predecessors. The biggest change concerns the widening of the jurisdiction of the parties concerned, for example, in relation to the current practice in the water rights courts. Special attention has to be given to the content and comprehensibility of information intended for the general public.

CHAPTER 4

Effluent loadings from the forest industry

1	**Measurement of effluent discharges**	**31**
2	**Discharge levels from manufacturing processes**	**32**
3	**Process modifications to reduce emissions**	**34**
3.1	Sulfate pulp production	35
3.2	Mechanical pulps	39
3.3	Recycled fiber	40
3.4	Paper and board	40
3.5	Sawmills	40
	References	41

CHAPTER 4

Effluent loadings from the forest industry

1 Measurement of effluent discharges

Forest industry effluents contain:

- Wood, either in its original form or in greatly changed form
- "Wood polymers"
- Process chemicals and their reaction products
- Various fillers and auxiliary chemicals.

The composition varies considerably.

The most common measurements used for characterizing these effluents are:

- BOD_7
- COD_{Cr}
- suspended solids (GF/A filter)
- phosphorus
- nitrogen
- AOX
- Cloro-organics
- Color
- Toxicity

In terms of their chemical composition, effluents are still not completely understood, particularly those arising from chemical pulp production. In the case of bleaching effluents, for example, only 10%–30% of the compounds present have so far been identified. This naturally causes difficulty in making a proper assessment of the effects of such effluents and to choose the most appropriate methods of treatment. Attempts to characterize effluents have normally involved methods of measurement developed to serve the needs of effluent treatment technology. While this practice is basically sound, it has obvious shortcomings. Whatever method of measurement is used, mention should be made of the analytical method (standard) concerned. Despite progress in standardization, comparisons between one country and another are still difficult.

In Finland, biological oxygen demand (BOD) is measured and quoted as BOD_7. This means the amount of oxygen required by microbes to break down wastes over a period of seven days. Outside the Nordic countries, a five-day BOD is widely used. In special cases, BOD may be measured over periods of 20 or 24 days.

Chemical oxygen demand (COD_{Cr}) is a measure of the amount of oxygen consumed in the chemical decomposition of waste. COD_{Cr} reflects more than just the amount of organic matter present and should not be used as the sole measure of "organic" matter content as often seems to be the case. Separate analytical methods – basically evaporation and measurement of ignition residues – are available for this purpose. Potassium permanganate used to be employed as the oxidizing agent, and the subscript was therefore Mn. This agent is no longer used, mainly because of the incompleteness of oxidation.

In Scandinavia, the suspended solids content of effluent is frequently determined using a GF/A filter with a nominal pore size of 1.6 µm. Filters with other pore sizes are also used. Pore size has a major influence on the results obtained, especially in the case of forest industry effluents because of their high colloidal content. Nutrients such as nitrogen are normally determined using the Kjeldahl method.

AOX is a measure of the halogens present in organic matter; in the case of forest industry effluents, the halogen is almost entirely chlorine. Finnish standards (SFS) have been published for all the methods of determination listed so far. However, other methods are used, both in Finland and elsewhere, which means the results are not always comparable (see also Appendix 5).

The above effluent parameters have been chosen because of their correlation with the impact on the receiving water. This is true at least in the case of BOD, nitrogen, and phosphorus determinations. Other important parameters used in characterizing effluents include color and toxicity.

2 Discharge levels from manufacturing processes

Effluent loadings depend on the process conditions and process technology employed. Fig. 1 shows how the BOD load from mechanical pulping effluents depends on pulp yield, while Fig. 2 demonstrates the dependence of AOX load from pulp bleaching on the use of elemental chlorine.

Effluent loadings vary with both the production process and the raw materials, even when the products being made are similar. The discharge values presented in the following table are therefore merely indicative of the magnitudes involved.

Effluent loadings from the forest industry

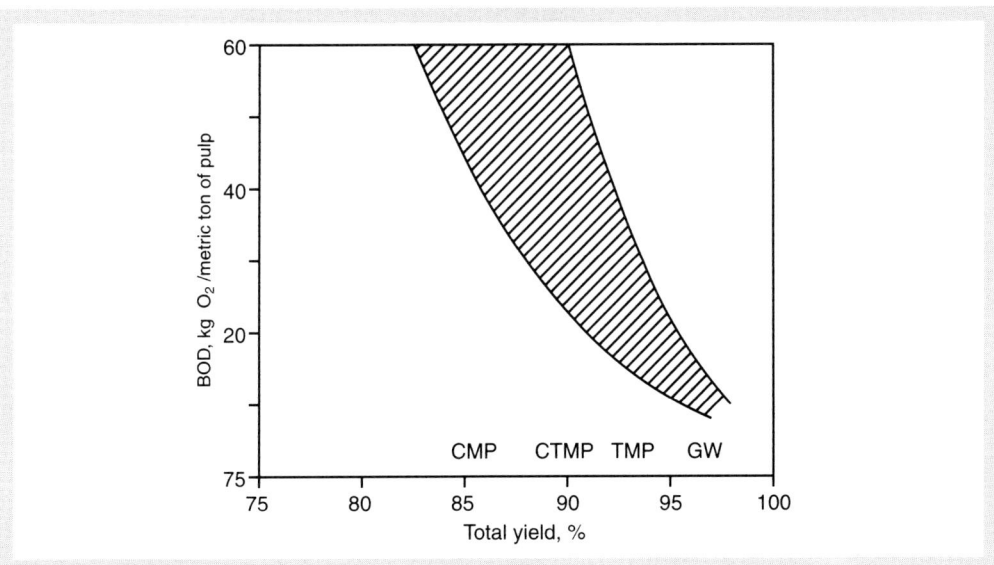

Figure 1. BOD load as a function of yield in mechanical pulping.

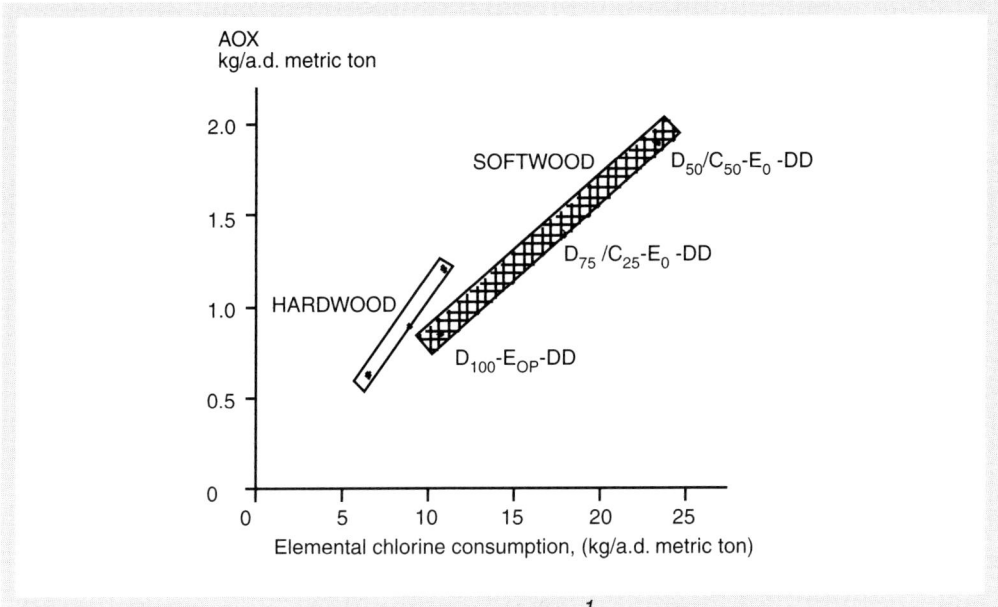

Figure 2. AOX load as a function of elemental chlorine consumption [1].

CHAPTER 4

Table 1. Effluent loads from the manufacture of forest products.

Product	Effluent cm³/metric ton	Suspended solids, kg/metric ton	BOD kg/metric ton	COD kg/metric ton	N g/metric ton	P g/metric ton
Pulp						
Sulphate pulp, unbleached	20–60	12–15	5–10	20–30	200–400	80
Sulphate pulp, conventional bleaching[a]	60–100	12–18	18–25	60–120	300–500	120
Sulphate pulp, ECF or TCF[b]	30–50	10–15	14–18	25–40	400–600	100
Sulphite pulp, conventional bleaching	150–200	20–40	30–40	60–100	100–200	60
Groundwood, unbleached	6–10	10–30	10–15	30–50	100–200	50
TMP, unbleached	6–15	10–30	15–25	40–80	100–200	70
TMP, peroxide bleached	6–15	10–30	20–40	60–100	200–300	100
Recycled fiber, deinked	10–20	5–10	20–40	40–80	100–200	40
Paper and paperboard						
Fine paper, coated	30–50	10–20	3–8	10–20	50–100	5
Newsprint	10–25	5–10	1–3	2–4	10–20	5
Folding boxboard	10–25	5–10	2–4	3–6	50–100	8
Sack paper	15–30	5–10	2–4	4–8	100–200	15
Tissue	20–40	5–10	1–3	3–6	50–80	8

The figures in the table were compiled from measurements made at Finnish mills.
[a]Kappa number 20–30, depending on wood species
[b]Extended cooking, oxygen delignification, kappa number 8–14, depending on wood species

AOX-levels for sulphate pulping with conventional bleaching were between 2 and 5 kgCl/t pulp and with ECF-bleaching 0,4-1,0 kgCl/t pulp.

3 Process modifications to reduce emissions

Figs. 3 and 4 schematically show the usual procedures adopted in seeking to reduce effluent discharges. Accurate information on the way the process operates is naturally needed before process modifications can be made. In most cases, anticipated future changes in production, both qualitative and quantitative, should be predicted as accurately as possible.

The following is a brief account of the process modifications recently introduced or currently being introduced in the interests of environmental protection and,

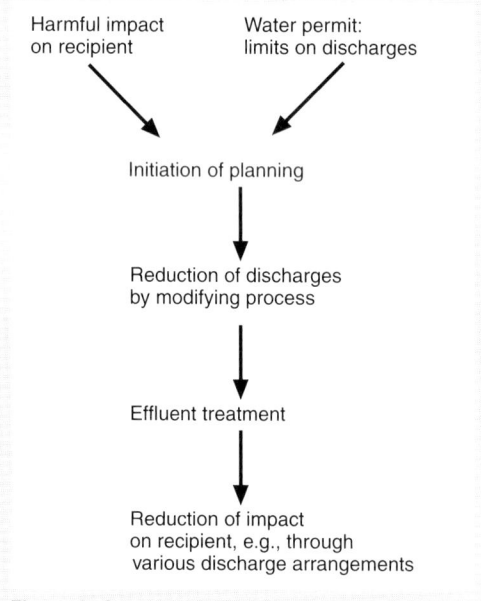

Figure 3. Sequence of actions for reduction of effluent discharges.

34

in particular, water pollution control. Details of these modifications can be found in other books in this series.Sulfate pulp production.

3.1 Sulfate pulp production

The sulfate process is the main method for the production of chemical pulp, and will probably continue to be for some time (see production forecasts, Appendix 3). The driving force behind the development of sulfate pulp production (see Fig. 5), particularly in the past ten years, has been the desire to reduce or discontinue completely the use of chlorine chemicals to bleach the pulp. This in turn means reducing the amount of lignin remaining in the pulp before the bleaching stage. The consumption of bleaching chemicals is directly proportional to the pulp's lignin content, as reflected by its kappa number (see Fig. 6). As bleaching is almost the only stage in the process where wastes enter the water leaving the process, the final effluent loading is also directly proportional to the lignin content of the pulp after cooking (see Fig. 7).

The principal methods currently used to achieve lower pulp kappa numbers are:

- Extended cooking
- Oxygen delignification
- Alkaline extraction of the pulp.

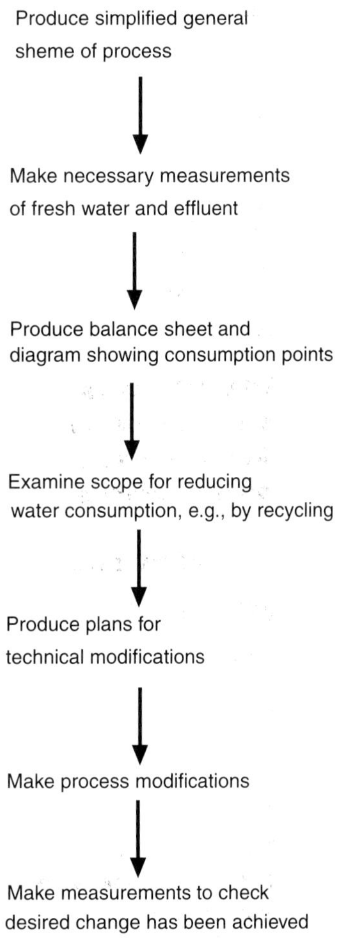

Produce simplified general sheme of process
↓
Make necessary measurements of fresh water and effluent
↓
Produce balance sheet and diagram showing consumption points
↓
Examine scope for reducing water consumption, e.g., by recycling
↓
Produce plans for technical modifications
↓
Make process modifications
↓
Make measurements to check desired change has been achieved

Figure 4. Sequence of actions for making process modifications.

In extended cooking, a high concentration of active alkali is maintained throughout the cook using a suitable dosing scheme. Oxygen delignification refers to the removal of lignin by treating the cooked pulp with oxygen at alkaline pH. The spent liquor from this stage is concentrated by evaporation and then burned.

Alkaline extraction provides a way of removing residual lignin from the pulp at high pH. This method is currently in the testing stage, while the other methods just mentioned are already in general use.

CHAPTER 4

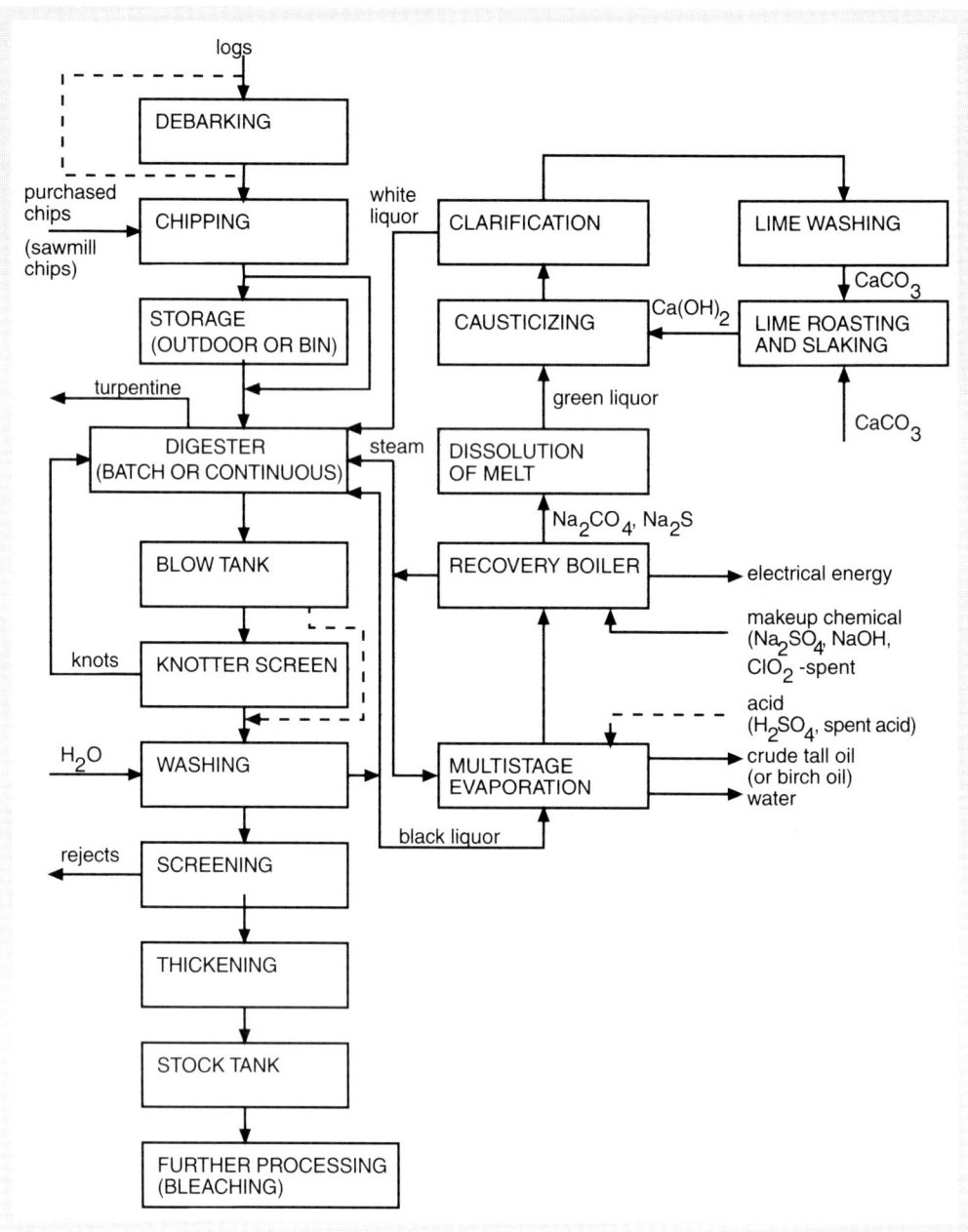

Figure 5. Scheme for production of sulfate pulp.

Effluent loadings from the forest industry

Processes that make use of chlorine dioxide and oxygen compounds such as peroxide and ozone have gradually replaced bleaching processes employing chlorine and chlorine dioxide. This shift away from chlorine-containing bleaching agents has reduced AOX loadings. COD loadings, which reflect the organic content of effluents, have also declined in relation to pulp kappa number. On the other hand, the introduction of oxygen-based bleaching agents does not seem to have removed the acute toxicity of the process discharges. Toxic compounds such as fatty acids and resin acids do not decompose and therefore pass into the effluent.

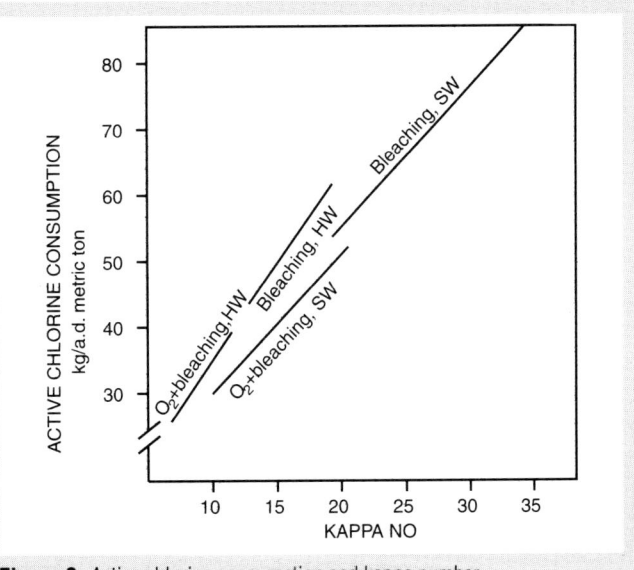

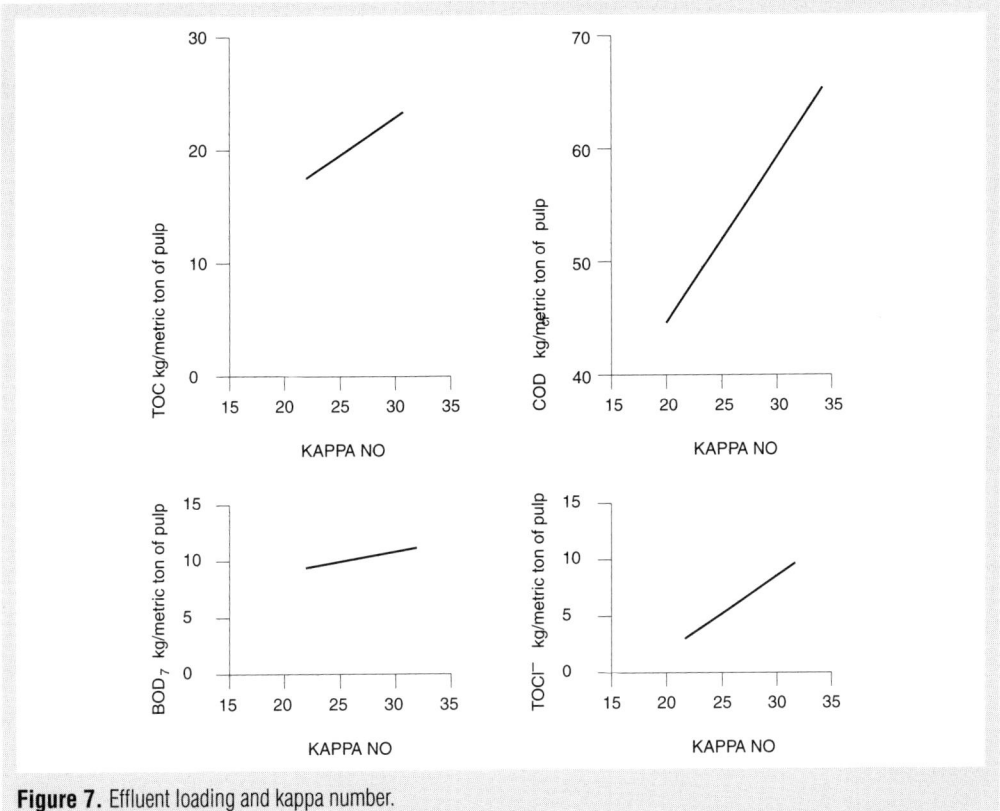

Figure 6. Active chlorine consumption and kappa number.

Figure 7. Effluent loading and kappa number.

CHAPTER 4

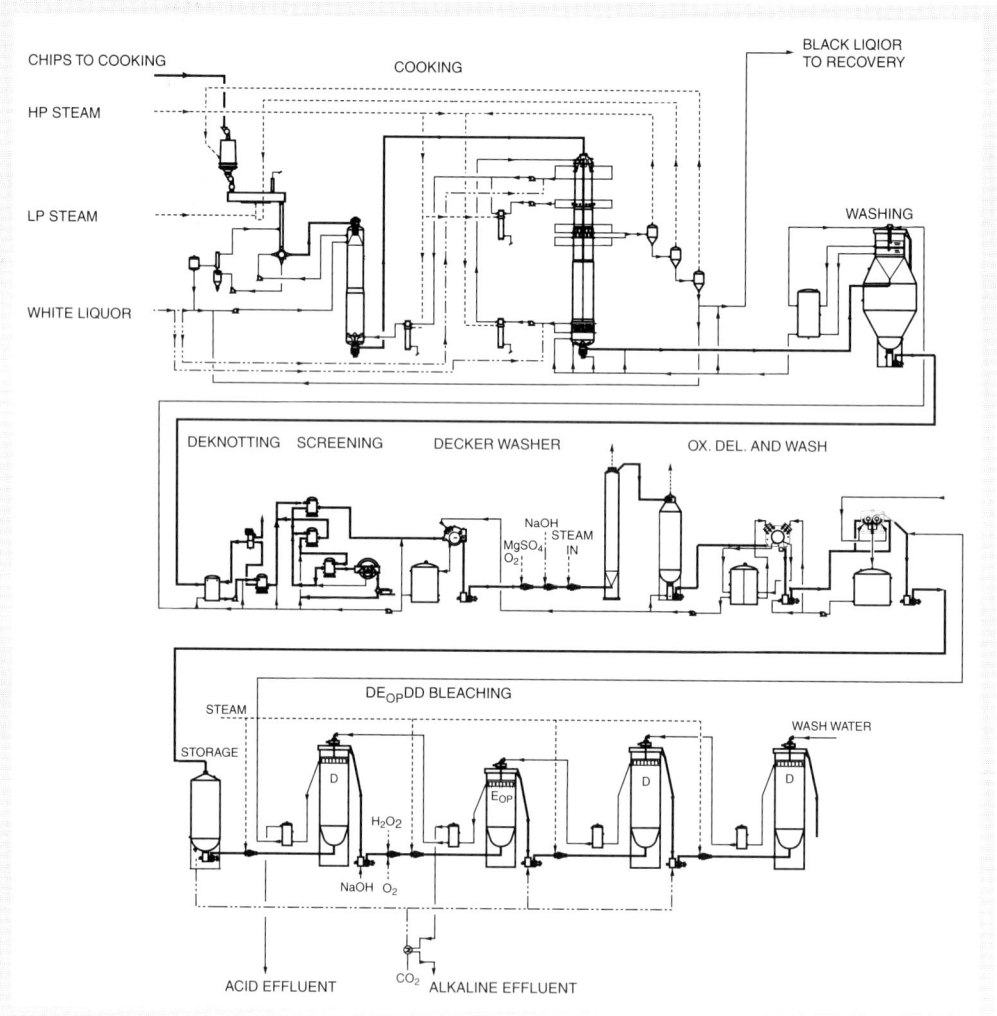

Figure 8. Schematic representation of a modern sulfate pulp mill.

Other important changes in sulfate pulp production include the changeover to dry debarking of the wood raw material and evaporation of black to higher dry solids contents (65% dry solids or more). The first has reduced effluent loadings and the second sulfur emissions to the air. It has also affected the chemicals balance, which in turn has meant using almost exclusively sodium compounds as makeup chemicals.

How the sulfate pulping process develops in the near future depends on what changes are made to the bleaching process. A schematic representation of a modern sulfate process is presented in Fig. 8. In this scheme, some of the spent bleaching liquors are taken for evaporation and burning. Despite this, the effluent load still depends on the effectiveness of wastewater treatment, and is likely to do so for some time to come.

3.2 Mechanical pulps

In terms of effluent loadings from mechanical pulp production, the most important factor has proved to be pulp yield (Fig. 1) . Pulp yield depends very much on whether the pulp is bleached and, if so, by what means. The biggest cause of loadings is peroxide bleaching.

In Finland, the main types of mechanical pulp produced are groundwood, pressure groundwood, and thermomechanical pulp. None of these poses particular problems in relation to effluent loadings. Most of the effluent loading is removed by mechanical treatment, and the rest is removed almost entirely by biological treatment.

However, even after treatment, the effluent is not entirely without effect on the receiving water. It contains harmful substances originating from the wood often bound to very small solid particles, as well as nutrients.

No particular process improvements that would help to reduce effluent or other environmental loadings are at present discernible. Perhaps the most notable step forward recently has been the changeover to dry debarking of the wood.

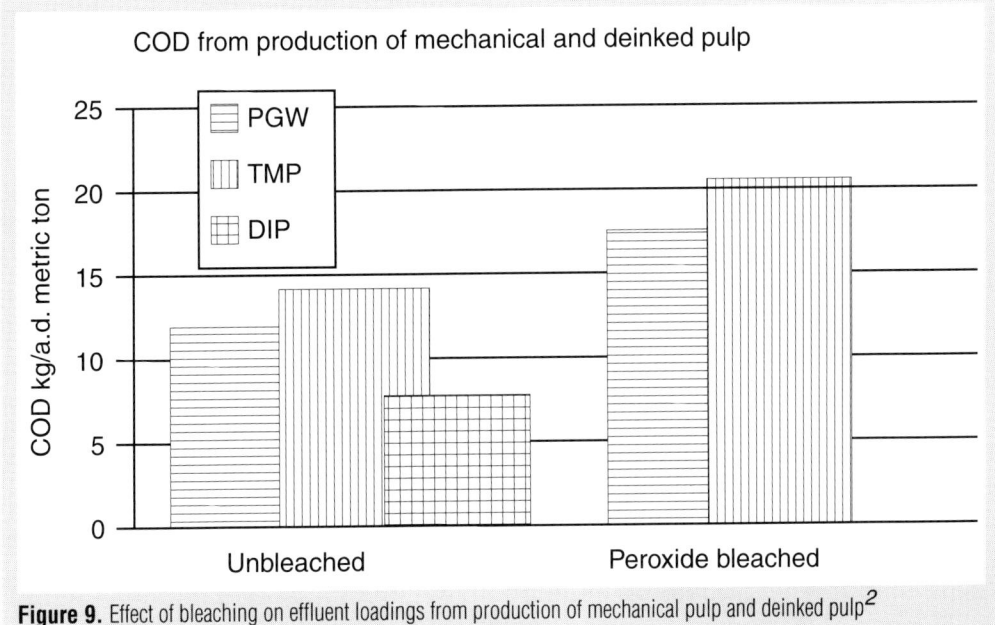

Figure 9. Effect of bleaching on effluent loadings from production of mechanical pulp and deinked pulp[2]

Turning to the treatment of process wastewaters, one significant step forward has been the partially successful changeover to evaporating the effluents and burning the concentrate. Other approaches have been tested, such as freezing the effluent and recycling the clean water obtained.

CHAPTER 4

3.3 Recycled fiber

Fiber recycling results in substantial process losses, partly as solid waste and partly as matter dissolved in process effluents. As fiber recycling becomes more widespread as a method of producing pulp, sludge handling problems are becoming more acute. As regards process development, the situation is similar to that in mechanical pulping, i.e., little has been reported and little new has been introduced. Fig. 9 illustrates the role played by bleaching in increasing effluent loadings.

3.4 Paper and board

In attempts to reduce the effluent loadings from paper and board manufacture, detailed analysis of the way the process functions has always played a key role. This means producing water and pulp balances and using these to rationalize the process and at the same time keep down effluent loadings. Areas where improvements can generally be achieved are:

- The cooling of uncontaminated waters and the re-use or separate sewerage of sealing water
- Recycling and, if necessary, cooling of water from vacuum pumps
- Discharging waste process water at only a few points (preferably only one) and even then after it has been treated.

In papermaking, it seems that once water consumption has been reduced to certain levels, a number of effluent parameters start to require more attention. In the manufacture of mechanical printing paper, for example, efforts to lower water consumption to 10–12 m^3/t lead to solids removal taking priority in the treatment of the circulating waters. When the aim is 4–8 m^3/t, attention also has to be given to the removal of dissolved organic matter. Lowering water consumption to 2–4 m^3/t requires action to manage salt concentrations.

The results depend, among other things, on the treatment process used. Biological treatment removes dissolved organic matter. However, recycling the biologically treated water would appear to be restricted by the resulting reduction in product brightness. This also calls for a more effective way of removing solids than conventional flotation and sand filtration. Taking costs into account, a suitable level of solids removal is often that achieved using residual microbes.

Treating the circulating water using methods such as nano- and ultrafiltration removes very small solid particles as well as some dissolved high molecular mass material. Achieving good results here demands close attention be given to management of surface chemistry during the manufacturing process.

3.5 Sawmills

At sawmills, effluent problems have normally been associated with log floating and wetting in the log storage area. The surface treatment of sawn timber has also caused some problems. In view of the special nature of the problems at sawmills, readers should refer to the relevant literature.

References

1. Soteland, N., Svensk Papperstid. (17):43 (1987).
2. Malinen, R., Wartiovaara, I., Välttilä, O., Scenario analysis of possible courses of development up to the year 2010, Sytyke 22 (in Finnish), Helsinki, 1993, pp. 85.

CHAPTER 5

Raw water treatment

1	**Quality requirements**	43
2	**Impurities**	43
3	**Treatment methods and equipment**	44
4	**Water supply system**	53
	References	55

CHAPTER 5

Raw water treatment

1 Quality requirements

The large number of different production processes employed in the wood-processing industry and the production problems encountered at individual mills mean that the requirements for water quality vary from one case to another. The raw waters used in these processes therefore vary considerably. Water quality, of course, can be improved through a range of treatment measures or by compensating for certain undesirable properties of the water in the process itself. Standards and test methods such as those issued by TAPPI[1] contain guidelines for the types of water needed to manufacture certain products.

In Finland, the main requirements for water quality concern color and iron and manganese concentrations.

2 Impurities

The main impurities that need to be removed from raw water are the following[1]:

Color. In Finland, surface waters often contain humus, which gives the water a brownish or yellowish color. Color is measured in Pt units. Color is assessed by comparing the Pt value of the sample with the values of standard solutions (for analyzing methods see Appendix 5).

Turbidity. Surface water often contains very small particles of suspended organic and inorganic substances, which make the water turbid. Turbidity usually is measured by comparison with solutions containing known concentrations of silicon dioxide (SiO_2). A method based on the back-scattering of light also is used.

Hardness. The hardness of water is due almost entirely to the presence of dissolved calcium and magnesium salts. In Finland, most natural waters are soft. Hardness is measured using the German system in degrees of hardness (°dH). One degree of hardness corresponds to 10 mg/l of dissolved calcium and magnesium salts calculated as calcium oxide.

Alkalinity. The alkalinity of water depends on its content of carbonates, bicarbonates, and hydroxides. Alkalinity is measured in milli-equivalents per liter (meqv/l).

Iron and manganese. These elements are present in water as a range of compounds in which they have different oxidation states. In surface waters, iron is invariably present as a colloidal form of hydrated iron (III) oxide, although the iron (II) form may also be present. In groundwaters, iron is usually found as iron (II) bicarbonate. Chemical pulp fibers absorb iron, causing the pulp or paper to appear slightly yellowish. Iron compounds also can form flocs, causing paper to exhibit small flecks of color. The effects of manganese are similar to those of iron. However, during bleaching, manganese is oxi-

CHAPTER 5

dized to permanganate, which imparts color to the pulp. Manganese also can form other colored compounds.

Free carbon dioxide and oxygen. At low pH, these dissolved gases can cause considerable corrosion.

Chlorides. At high concentrations and under certain conditions, chlorides also can cause corrosion, although they do not normally interfere with the papermaking process. Their presence may be critical, however, in the manufacture of certain specialities such as insulating papers.

3 Treatment methods and equipment

Choice of treatment process

In industry, water treatment is usually carried out:

- To remove solids
- To remove color and organic substances
- To remove iron and manganese
- To counteract hardness and remove all dissolved salts
- To disinfect the water.

If the water is to be used for drinking purposes at the mill, odor and taste may have to be considered, too. With the exception of solids removal, all of the above require some form of chemical treatment. This usually consists of the following unit processes:

- Passing the water through screens and sieves
- Addition of appropriate chemicals
- Rapid mixing
- Flocculation (with addition of coagulating agents if necessary)
- Settling (clarification)
- Filtration
- Further addition of chemicals.

The effects on water quality of these different unit processes are shown in Table 1. The treatment of water for boiler plants is a separate issue, and is discussed under ion exchange.

Table 1. Effects of different unit processes on water quality.

Property	Aeration	Coagulation and clarification	Slow filtration (alone)	Fast filtration (after coagulation and clarification)	Disinfection (with chlorine)
Bacteria	0	+	+++	+++	+++
Color	0	++	+	+++	0
Turbidity	0	++	+++		0
Taste and odor	+	(+)	+	(+)	+++
Corrosive properties	+++	(-)	0	(-)	0
Iron and manganese	++	++	+++	+++	0

Other factors that influence the final choice of treatment method and equipment include the volume of water to be treated, the space available for a raw water treatment plant, and to some extent how well the running and supervision of the plant can be integrated with the mill's other operations.

Mechanical treatment

The mechanical treatment performed before chemical treatment usually requires:

- A water intake system such as pipeline and raw water pumping station
- Screens and filters
- A treatment plant and pumping station for the mechanically treated water.

All raw water used by Finnish pulp and paper mills is surface water. The water intake station is located close to the watercourse with the abstraction point suitably situated where the water flow is slow. This minimizes the amount of coarse solid particles removed with the water. If the water flow is fast, reservoirs can be built for the purpose of water abstraction. There are often other good reasons for building such reservoirs.

Water is usually abstracted through sub-surface suction pipes which may be fitted with their own filters. The rate of flow of water through suction pipes is only around 0.8–1.0 m/s. Smaller pipes (diameter up to 600 mm) are usually of polyethylene, while larger diameter pipes are made of polyethylene or wood. The pipes are weighted, the weight usually being 25%–50% of the upthrust on the empty pipe.

The total area of the holes in the filter built onto the end of the water intake pipe should be at least three times that of the cross-sectional area of the pipe. The individual holes in the filter should have a diameter of 15–20 mm.

Mechanical treatment normally begins with a screen, which in most cases is a fairly fine one. The total free space is dimensioned for water flows of 0.2–0.5 m/s, where 0.2 m/s

CHAPTER 5

applies to rivers where frazil ice and debris is a problem, while 0.5 m/s is used for water abstraction from a reservoir. The recommended bar spacing is 10–25 mm. The effect of the screen bars and trash content of the water are taken into account in dimensioning the screen. The following equation can be used for the calculation:

$$A = (Q/v) \times k_1 k_2 k_3 \tag{1}$$

where A is Total surface area of holes (m²)
- Q Design flow rate (m³/s)
- v Water flow rate through screen openings (0.2–0.5 m/s)
- k_1 Shape factor (1.1 for round bars and 1.25 for rectangular bars)
- k_2 Screen reduction factor due to bars: $k_2 = (a+d)/a$
  where d is Reduction in area due to bars
    - a Free area
- k_3 Clogging factor (~ 1.25).

The screen should be cleaned from time to time, although – for plants dealing with less than 1.0 m³/s – this may not be necessary.

Major factors affecting the choice of pump for pumping stations include the volumes of water involved and repair and servicing considerations. Smaller stations use vertical submersible pumps, whereas larger stations rely more on dry-installed horizontal centrifugal pumps. Stations are equipped with reserve pumps as necessary. More information on the criteria for pump choice will be discussed later.

Mechanical treatment proper begins with straining to remove small organisms, leaves, grass, and other trash. Strainers are divided into two groups depending on their mesh size:

- Conventional strainers, mesh size over 0.1 mm
- Micro-strainers, mesh size under 0.1 mm.

Strainers can be of the drum type or they can be flat. For cleaning, both have to be flushed with water under pressure, either continuously or from time to time as dictated by the pressure drop.

The choice of mesh size depends on the material that needs to be removed. In order to remove algae, for example, a mesh size of less than 100 m (usually 30–60 m) is needed.

The water passing through the strainer collects in a storage tank or pumping tank. The tank design normally allows a storage capacity of 15–30 minutes. From here, the water is pumped into the distribution system, where the pressure is around 0.3 MPa. Booster pumps can be used if a higher pressure is required.

Chemical treatment
Treatment chemicals

In Finland, raw water is often colored, as turbid surface water may contain humus as well as iron and manganese, either in oxidized form or bound to organic matter. Water of this type is treated using the following scheme:

Aeration + coagulation + addition of alkali + clarification + sand filtration.

As far as the pulp and paper industry is concerned, the biggest problems are color and iron and manganese contents. Several methods, including the scheme mentioned above, can be used for their removal. The removal of color and organic compounds usually requires a method based on chemical precipitation.

Some reduction in organic matter content, of course, can be achieved by simple filtration.

The removal of iron and manganese, on the other hand, usually requires:

1. Oxidation of iron (II) and manganese (II) to oxidation state III

2. Hydrolysis of the trivalent cation to produce the hydrated hydroxide

3. Coagulation of the hydrated hydroxide

4. Removal of the coagulate.

Several methods and chemicals are available for oxidation and coagulate removal.

The first step in the chemical treatment process is usually aeration, although in the pulp and paper industry this is not much used. Aeration is designed to achieve the following:

- Oxidation of iron and manganese ions

- Reduction in carbon dioxide content and hence in risk of corrosion

- Removal of malodorous components such as methane and hydrogen sulfide, which improves taste and odor.

Treating the water with an oxidizing agent such as chlorine or permanganate can accomplish oxidation.

The next step is usually chemically-induced coagulation and flocculation. This means using a chemical reaction to produce flocs that will form a sediment. The sequence of events is addition of chemical, rapid mixing, and finally gentler stirring. The addition of chemicals produces micro-flocs (coagulate), which forms larger flocs as the mixture is gently stirred.

Tests are made first to determine the amounts of coagulants that need to be added.

CHAPTER 5

One such commonly used chemical is aluminium sulfate. A suitable rate of addition can be calculated from measurements of zeta potential as follows:

- Plot a zeta potential curve as a function of aluminium sulfate addition.
- Use the curve to determine the addition that corresponds to a zeta potential of −7 to −10 mV.
- Adjust the zeta potential as close as possible to ±0 by adding a cationic polyelectrolyte.

Aluminium sulfate reacts with the alkaline compounds naturally present in the water and also with any added lime or sodium carbonate according to the following general reactions:

1. Natural alkalinity

$$Al_2(SO_4)_3 + 3Ca(HCO_3)_2 \rightleftarrows 2Al(OH)_3 + 3CaSO_4 + 6CO_2 \qquad (2)$$

2. Lime

$$Al_2(SO_4)_3 + 3Ca(OH)_2 \rightleftarrows 2Al(OH)_3 + 3CaSO_4 \qquad (3)$$

3. Sodium carbonate

$$Al_2(SO_4)_3 + 3Na_2CO_3 + 3H_2O \rightleftarrows 2Al(OH)_3 + 3Na_2SO_4 + 3CO_2 \qquad (4)$$

Although these equations do not give a complete picture of the coagulation reactions, they can nevertheless be used to calculate the theoretical amounts of the various chemicals needed.

Another commonly used coagulant is sodium aluminate. This is normally employed as a supplementary coagulant to ensure proper treatment of cold waters or to coagulate residual aluminium sulfate. In the latter case, the reaction is:

$$3Na_2Al_2SO_4 + Al_2SO_4 + 12H_2O \rightleftarrows 8Al(OH)_3 + 3Na_2SO_4 \qquad (5)$$

Supplementary coagulants are used in cases where floc formation would otherwise be unsatisfactory. The most common such agents are activated silicic acid, certain naturally occurring organic compounds and synthetic polyelectrolytes.

Other chemicals used, mainly for pH adjustment, include certain calcium and sodium compounds. As raw water in Finland is generally soft, addition of lime does not increase the hardness of water unduly. At small water treatment plants, water storage and the addition of chemicals could justify the use of sodium carbonate and sodium hydroxide.

Raw water treatment

Rapid mixing

The mixing of chemicals should take place uniformly and rapidly. This is achieved using mechanical mixers, turbulent flows in the water, or mixing in raw water pumps.

Numerous different type of mechanical mixer are in use. These are dimensioned for the following:

- Mixing times of 0.2–2 min
- Peripheral speeds of 0.6–2 m/s
- Energy consumption of 1.0–4 Wh/m^3.

By feeding the chemicals into the intake pipe of the raw water pump, rapid mixing is simple and takes place inside the pump chamber. Feeding the chemicals into a pressure pipe is also an option, in which case baffles mounted inside the pipe enable mixing.

When flow turbulence is used to bring about mixing, a drop of 0.2–0.8 m achieves the necessary turbulence.

Flocculation

Rapid mixing results in the formation of micro-flocs. The water now has to be stirred in such a way that these micro-flocs combine to produce larger flocs that will settle to the bottom. Either flow turbulence or mechanical stirring can achieve this. In Finland, mechanical stirring is used almost exclusively.

Mechanical mixers are either vertical or horizontal. The value of the velocity gradient for a particular mixer is calculated from the equation:

$$G = \sqrt{C_o A v^3 / 2 V \vartheta} \qquad (6)$$

where A is Area of vanes (m^2)
v Velocity of vanes relative to that of the water (usually 3/4 of absolute velocity of vanes)
V Volume of flocculation basin (m^3)
ϑ Kinematic viscosity of water (m^2/s)
C_o Specific resistance constant of mixer (1.8 for straight, smooth vanes).

G should have a value of about 100 at the start of flocculation and of about 10 at the end.

Other parameters of mechanical stirrers usually vary within the following limits:

- Retention time, t, 30–60 min
- Input, Gt, 10^4–10^5
- Initial peripheral speed 0.4–0.8 m/s
- Final peripheral speed 0.1–0.3 m/s
- Area of vanes 10%–25% of corresponding area of water impinging at right angles

CHAPTER 5

Stirrers should be designed to achieve the maximum number of collisions between the growing flocs and the colloidal matter in the raw water.

Clarification

Most of the flocs (usually around 90%) formed during coagulation and stirring are removed during clarification. This takes place either by sedimentation or by flotation. The methods used can be classified as follows:

- Horizontal clarification
- Vertical clarification
- Flotation
- Other methods.

The most frequently used are flotation and horizontal clarification, and these are discussed in more detail below.

In horizontal clarifiers, the water flows through the tank in the horizontal direction while the particles settle downward perpendicular to the flow. The tanks used are either rectangular or circular in cross-section and are fitted with one or several bases. Water is fed into the tank to impinge on a damping wall at a rate not exceeding 0.3–0.45 m/s. At least two tanks are used, so that one can be cleaned out periodically.

Tanks are dimensioned as follows:

- Surface load 0.75–1.5 m/h (usually 1.0 m/h)
- Depth 3–4 m
- Length 3–6 times depth (usually 4–5 times)
- Reynolds number kept below 10 000 if possible
- Froude number over 10^{-5}
- Inclination of bottom 1:20–1:100
- Retention time 2–4 h
- Effective volume 75% of total volume.

It is the surface load that usually determines the tank's dimensions. The hydraulic characteristics produced by the tank are checked by calculating the Reynolds and Froude numbers. The rate at which the flocs settle also can be used as a basis for dimensioning.

The clarified water is led out of the tank, usually by means of adjustable collector channels. The edge load is usually 5–10 m^3/m/h and the velocity in the channels is 0.2–0.3 m/s.

In flotation, the solid particles are lifted by means of air bubbles to the surface, where they are either removed by skimming or collected as an overflow by raising the level from time to time. Air bubbles are introduced into the tank as dispersion water in which air has first been dissolved in a pressurized tank. The advantages of flotation are that it takes up relatively little space and is very effective with soft, humus-containing waters.

The following design parameters are used:

- Surface load 4–10 m/h, usually 5 m/h
- Tank depth 2–3 m
- Retention time 15–30 min
- Dispersion water requirement about 10% of raw water feed
- Pressure in dispersion tank 0.5–0.8 MPa.

Other clarification systems include lamella and tubular clarifiers (Fig. 6.1). Lamellae or tubes, which render the flow laminar, have sometimes been installed in old clarifiers to increase their capacity.

Filtration

In the production of fresh water for the pulp and paper industry, the flocs that have passed through with the clarified water from chemical treatment are removed by filtration.

Filter design and capacity are determined by the maximum permitted pressure drop and the requirements for water quality.

Filters employing a filter medium may have filter beds with one, two, or more layers. Single-layer filters consist of sand particles with a size of 0.4–1.2 mm.

In multilayer filters, the filter bed is made up of layers of different materials with different particle sizes. The rate of filtration is usually greater than that achieved with a single-layer filter. In the case of a single-layer filter, the sand lies on a base, below which there is a space for collection of filtered water and from where the bed can be flushed.

The rate of filtration is normally around 5–7 $m^3/m^2/h$. The flow of water through the base is arranged with the help of various openings or jets.

The filter bed is flushed clean with water, or a mixture of air and water, directed upward through the base. The water pressure needed is 0.08–0.1 MPa. Flushing is carried out until the sand has gained at least 30%–40% in volume (usually 50%). The volume of flushing water needed is 2%–4% of the volume of water filtered. A filtration resistance of 0.015–0.02 MPa generally indicates that the filter bed needs cleaning.

In addition to the above, use is also made of pressure filters and inverted filters, one version of which is the contact filter. Some filters are shown in Fig. 1.

CHAPTER 5

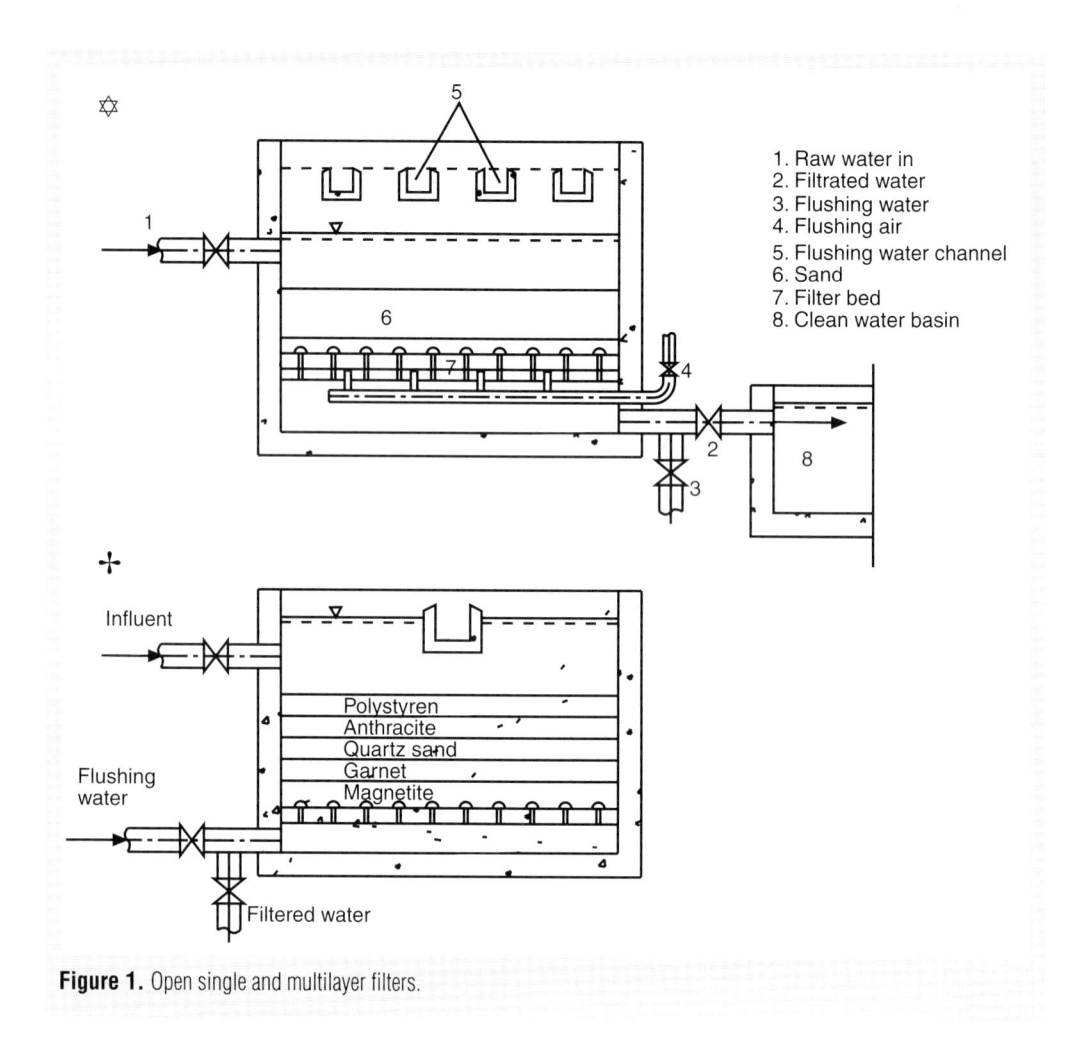

Figure 1. Open single and multilayer filters.

1. Raw water in
2. Filtrated water
3. Flushing water
4. Flushing air
5. Flushing water channel
6. Sand
7. Filter bed
8. Clean water basin

Increasing use is being made of pressure filters fitted with filter fabrics or meshes. While these can be used for filtering large volumes of water, their use is normally restricted to supplying water to individual items of equipment. Pressure filters are often used as the first stage in rendering paper mill circulating water suitable for new uses at the mill. Such filters generally require effective automatic washing arrangements if they are to function efficiently.

Disinfection

Before it is led into storage tanks or the supply network, the treated water is disinfected, especially if it is intended for drinking. The disinfectants used at water treatment plants are normally oxidizing agents such as chlorine, hypochlorite, and ozone. Other methods, including the use of UV radiation, have also been tested (see also Appendix 5).

The storage tanks used for chemically purified water are normally designed for a storage capacity of 4 hours.

Other treatment methods

Of the other methods used for water treatment at pulp and paper mills, the main one is ion exchange. Examples are treatment of condensate being returned to the mill's power plant and the production of power plant top-up water.

In most cases, condensates are first filtered to remove impurities originating from the pipelines. The filters used are designed to resist both heat and mechanical wear and tear. The filtered water is led to an ion exchange unit. These are usually mixed bed ion exchangers (containing both anionic and cationic exchange resins), possibly preceded by a cation exchanger.

In the case of plants producing top-up water, the water is first chemically treated before being passed through anionic, cationic, and mixed bed units.

Depending on the type of ion exchange resin used, the column is regenerated with acid, alkali, or sodium chloride solution. The washings are collected and neutralized before being discharged to the drain.

The volume of deionized water needed to flush the regenerated resin is normally 3%–10% of the volume of water treated. During washing, compressed air is used to increase the volume of the ion exchange resins.

In calculating the storage capacity for ion exchange-treated water, the fairly large volume of water needed for flushing must be taken into account as well as the interruptions to water production caused by regeneration.

4 Water supply system

The water supply system at a mill requires the building of pipelines to deliver different types of water.

The following types of water are generally available:

- Mechanically treated water
- Chemically treated water
- Drinking water
- Water for firefighting.

Water used for firefighting is usually chemically treated and delivered at a pressure of 1–1.2 MPa. If necessary, the system can also be supplied with raw water directly from a river or lake by means of diesel-powered pumps.

The pumps installed in the pumping station are either vertical or horizontal types dimensioned according to the amount of water needed. Vertical pumps are of three types, depending on the way they are employed: pumps installed in a well, submerged cantilever pumps, and submersible pumps. Factors to be considered when choosing and installing a pump include the type of watercourse, automatic operation, space requirement, and service and maintenance needs.

CHAPTER 5

Larger water treatment plants often use horizontal pumps, in which case the pump is positioned above the surface of the intake basin. The pumping station also can be built alongside the intake basin, but in this case flooding is a danger.

The pipelines are dimensioned on the basis of hydraulic calculations, for which various programs have been developed. The maximum recommended water velocities are:

- Intake pipes 1–1.5 m/s
- Pressure pipes 1.5–2.5 m/s
- Free flow pipes 1 m/s.

When designing the pipeline, sufficient space must be reserved for service and repairs. The choice of pipe material depends on the pressure requirements, the location, and the necessary corrosion resistance.

The valves in the main pipelines are slide and butterfly valves. Rapid closures must be avoided to prevent surges.

References

1. Virkola, N.E., Ed., *Production of Wood Pulp, Part II*, Turku, 1983, pp. 1376–1396.

CHAPTER 6

Effluent treatment

1	**Solids removal**	**58**
1.1	Clarification, flotation and filtration	58
1.2	Chemical coagulation and flocculation	67
2	**Biological methods**	**68**
2.1	Activated sludge process	68
2.2	Other aerobic treatments	81
2.3	Anaerobic processes	83
3	**Other treatment methods**	**86**
3.1	Activated carbon adsorption	86
3.2	Evaporation	87
3.3	Lignin removal process	88
3.4	Stripping	89
3.5	Ion exchange	89
3.6	Chemical oxidation	90
3.7	Freezing	90
4	**Removal of organic matter in effluent treatment**	**91**
	References	92
	Suggested reading	93

CHAPTER 6

Effluent treatment

General

Effluent loadings from the wood-processing industry can be reduced through process changes (internal approach) and by treatment once the effluent has left the process (external approach). In practice, the two approaches are combined in a way that yields the desired result at minimum cost. External treatment is frequently chosen because the necessary technology is available and its use does little to interfere with the production process itself. On the other hand, investments in external treatment are purely and simply expenditure items.

Internal process measures reduce loadings at source, and direct savings can be made in raw material costs (fiber, fillers, additives) and in energy consumption. In most cases, the necessary measures can be carried out by the plant's own personnel. However, changing over to a less-polluting process may result in considerable costs because of the more complex technology required. In spite of the costs, more emphasis is likely to be placed on internal process measures in the future.

Technological progress will continue via both internal and external measures and combination of these measures. Emissions will continue to diminish.

Effluents from the wood-processing industry contain wood material either in its original form or in some other form. The effluents also contain some chemicals used in the process, again in either original or modified form. Solids account for most of this material, but some is colloidal or present in solution. Effluents from wood-processing plants often have high contents of colloidal material. With a few exceptions, levels of nitrogen and phosphorus nutrients are low compared with, say, municipal wastewaters. The effluents also can be highly colored, notably those from chemical pulp production. Electrochemically, the species present are negatively charged, with zeta potentials usually between −60 and −10 mV. Some untreated effluents, for example those from the debarking of wood, are toxic to fish. Among the known toxic compounds present are fatty acids and resin acids. The effluent parameters normally used to classify effluents were presented in Chapter 4.

Broadly speaking, effluent treatment consists of equalization of variations in effluent flow and composition, followed by removal of solids and low molecular mass material (max. around 800 daltons). The latter also results in removal of some colloidal and dissolved high molecular mass material.

Physical-chemical treatment methods remove mainly solids and high molecular mass colloids, while biological methods and some physical methods such as distillation also remove low molecular mass compounds.

CHAPTER 6

1 Solids removal

1.1 Clarification, flotation and filtration

The principal methods used to remove solid matter from effluent are settling/clarification, flotation and straining/filtration. The method chosen depends on the characteristics of the solid matter to be removed and the requirements placed on the purity of the treated water[1].

Clarification (settling, sedimentation) relies on the particles of solid matter having a higher relative density than water. The effluent passes through a clarifier with a retention time of several hours. Solid matter settles to the bottom, from where it is removed. The most common type of external treatment used for forest industry effluents is mechanical clarification. The process can be divided into three stages: pretreatment, clarification, and sludge handling.

Pretreatment may consist of coarse screening, grit removal, neutralization, cooling, and equalization of flow. Clarification removes between 60 and 95% of solid matter. Finely suspended material that does not settle satisfactorily requires either mechanical flocculation or the addition of chemicals. The sludge obtained from clarification is usually thickened, dewatered, and then either used or disposed of.

Four types of settling take place in the clarifier:

- Discrete settling of individual particles
- Settling of flocs
- Hindered settling
- Thickening.

Discrete settling is unaffected by the presence of other solid particles. In this case, the particle has a characteristic settling rate that depends on its shape, its density relative to that of the fluid, and the viscosity of the fluid. This applies to solids such as grit. However, particles may sometimes join together to produce flocs, which have their own settling characteristics. Floc size and shape are constantly changing, which makes it virtually impossible to predict floc settling behavior mathematically. Settling rate is determined empirically. For settling to take place satisfactorily, the solids content should not exceed 1–1.5 g/l.

Hindered settling takes place largely during the final stage when the flocs form large agglomerates as they sink downward. The sludge concentration in such cases is higher than 1 g/l. Both empirical and theoretical methods are available for predicting this type of settling[2].

The final stage in settling is compaction thickening of the flocs at the bottom of the tank.

According to Stokes' law, the velocity of a spherical particle falling through a viscous medium is

$$V_s = (1/18/((\rho_p - \rho_1)/\mu))gd^2 \tag{1}$$

where V_s is the velocity of the particle, ρ_p and ρ_1 are the densities of the particle and the liquid, respectively, μ is the viscosity of the fluid, g is acceleration due to gravity, and d the diameter of the particle. It can be seen from this equation that a rise in temperature would increase the settling rate of the particles as both density and viscosity decrease with increasing temperature. In the sedimentation process, Stokes' law applies best to the settling of grit or similar particles.

Stokes' law is only valid for laminar flow, which is the type of flow in most clarifiers. According to Eq. 1 it is the particle diameter that has the greatest impact on settling rate.

A common type of primary clarifier is the solids contact clarifier, in which the effluent passes through a layer of sludge at the bottom of the tank. This filters out some solid particles, produces flocs, and thus accelerates settling. Greater efficiency can be obtained by employing slow stirring in the flocculation zone. Also used are earth basins and both vertical and horizontal clarifiers without contact sections.

Contact clarifiers have the following advantages over conventional types:

- Greater surface loads; shorter retention time in the clarifier allows smaller tanks to be used
- Use of a flocculation zone ensures uniform flow to the clarifier, i.e., there are no disruptive flows.

Some clarifiers featuring sludge recycling (Enso-Eimco HRC and YIT TKS) are shown in Fig. 1.

Clarifiers normally are designed according to surface load. Surface load refers to the flow of effluent in relation to the surface area of the clarifier.

$$A = k \times Q/v \tag{2}$$

where v is Settling rate of particles, m/h
        Q Flow rate, m³/h
        A Area of tank's clarification section, m²
        k Safety factor.

All particles with settling rates equal to or greater than the surface load remain in the tank. This requires laminar flow and free settling, i.e., an influent solids content of less than 1 g/l. The settling rate, v, is determined from laboratory tests, as are the properties of the settled sludge with a view to later handling.

CHAPTER 6

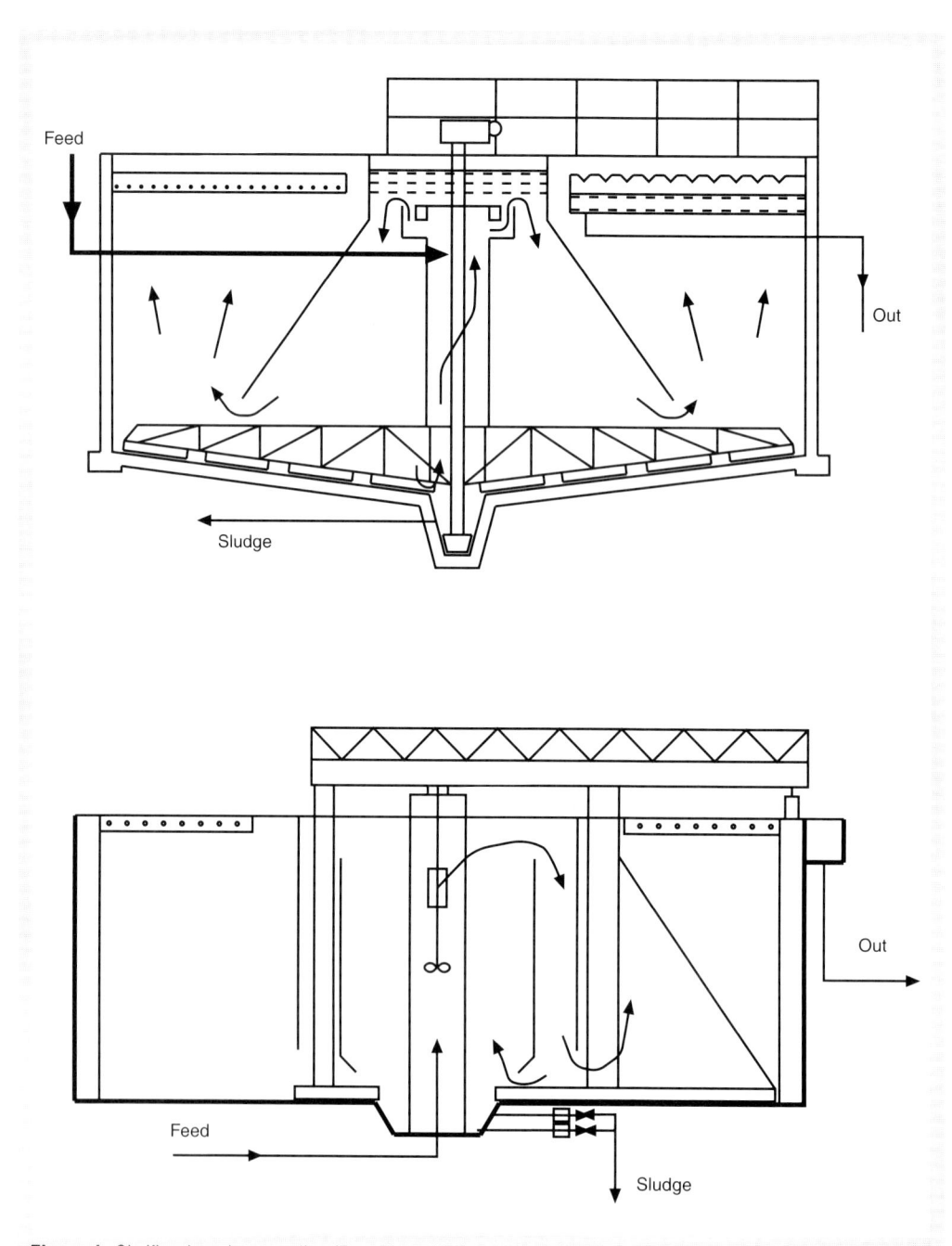

Figure 1. Clarifiers based on recycling (Enso Eimco HRC and YIT TRS).

In practice, within the clarifier there are both temperature differences and variations in flow conditions. Solids do not usually settle as discrete particles but rather as flocs. For this reason, tanks are often designed on the basis of laboratory tests and the use of correction factors.

The side load, R, determines the length of the outflow channel. Side load refers to the rate at which influent enters the tank in relation to the length of the side of the overflow weir. Effluent characteristics also have to be taken into account in designing weirs.

The surface load to the primary clarifier is usually 0.8–1.2 m/h, and the side load is something like 5–8 $m^3/(mh)$. In the biological treatment stage, the surface load to the secondary clarifier should be lower (normally 0.5–0.7 m/h).

Other important design factors include the water retention time and the bottom sludge load, i.e., the quantity of sludge settling to the bottom per hour. In primary clarifiers, the retention times are 3–6 h and the sludge loads are 50–150 kgm^2/h.

Clarifier performance is measured in terms of solids removal capacity. Solids reduction is the amount of solids removed in relation to the amount of solids entering the tank, and is usually given as a percentage.

The nature of the influent affects the functioning of the clarifier and the results obtained. Generally speaking, debarking effluent clarifiers and mixed effluent clarifiers are least efficient in removing solids. In the latter case, the effluent leaving the clarifier after the treatment of combined paper mill and debarking plant effluents usually has a fairly high solids content, despite the fact that the percentage reduction is relatively high.

The composition and flow of influent to the settling tanks usually vary considerably during normal mill operation.

The settling of very small particles is a delicate process, and even small changes in the conditions can result in these particles passing through with the water. The surface chemistry of such particles is highly dependent on pH, and pH therefore greatly affects the settling of this type of material. The settling of coarser particles is less affected by pH.

Clarifiers designed to remove fiber from paper and board mill effluents produce good results: The effluent leaving the clarifier has a low solids content, and solids removal rates can be as high as 90%. The surface charges and settling properties of particles usually can be improved by adding chemicals to aid coagulation and flocculation.

Variations in the temperature of the influent cause alterations to flow patterns and short-circuiting through changes in the density of the water in different parts of the clarifier. Short-circuiting interferes considerably with the settling of very small particles.

Problems have been caused at effluent treatment plants by sudden changes in effluent composition and flow rate, poor settling properties of the suspended solids, surface sludge, blockages, and damage to the equipment. Foaming is promoted by surface-active agents and also by air entrained in the effluent. The problem of surface sludge is largely confined to debarking plant clarifiers, where needles, bark, wood chips, and sawdust accumulate on the surface.

CHAPTER 6

The operation of a clarifier is monitored through automatic measurements and laboratory analyses.

The influent to the clarifier must be stabilized as much as possible in terms of both flow rate and composition by making adjustments to the process and through sewerage arrangements. The influent should be such that the solids readily settle out. Most detrimental to clarifier operation are surges in influent temperature and pH.

All control operations and modifications to clarifier tanks should be aimed primarily at eliminating short-circuiting. Accordingly, suitable values must be found for mixer speed, the height of cylinders, and the side load for the outflow. Regular cleaning of overflow weirs helps to reduce short-circuiting.

One modification of the conventional clarifier is the lamella clarifier (Fig. 2), which has been used for separate treatment of wastewater from coating color kitchens. This type of clarifier produces a laminar flow, which makes settling more efficient.

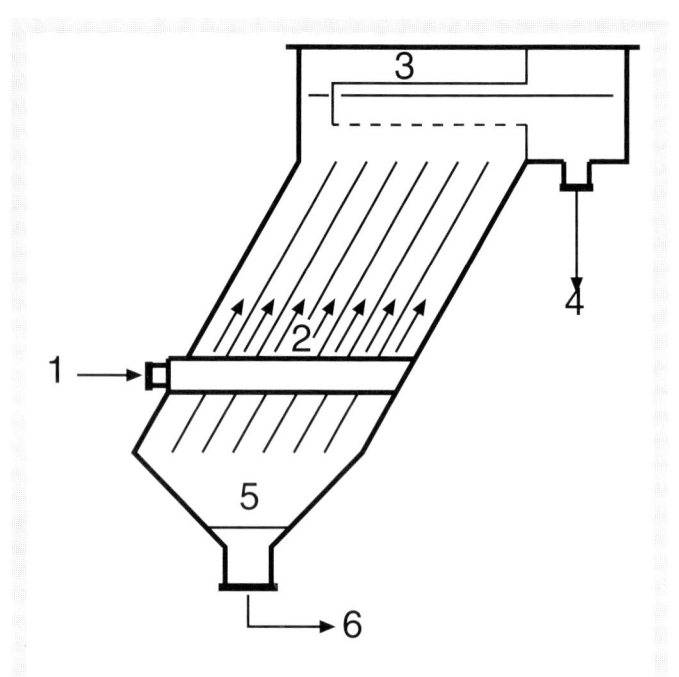

1. Effluent feed
2. Distribution section
3. Removal of clarified water
4. Supernatant out
5. Sludge collection
6. Sludge removal

Figure 2. A lamella clarifier.

The advantages of a lamella clarifier are its small space requirement and good performance; one disadvantage is the sliming up of the lamellae, which means increased maintenance.

Flotation clarifiers employ small air bubbles to carry suspended solids to the surface. The air is supplied by pumping into the clarifier a portion (5%–20%) of the clarified effluent or dispersion water made from fresh water.

Dispersion water is prepared by bringing water to a pressure of around 4–6 bar by means of a pump and passing air into it using an ejector. Air dissolves in the water under pressure. The dispersion water is then led into the first part of the clarifier and allowed to disperse through nozzles. As the pressure falls, tiny bubbles of air are released. As these rise toward the surface they become attached to solid flocs, which

Effluent treatment

they transport to the surface. The sludge forming on the surface is removed by skimming, either continuously or from time to time into a sludge run-off. The heaviest particles in the effluent being treated nevertheless will fall to the bottom, which therefore is fitted with its own sludge removal system.

Flotation is best suited for clarifying waters containing light sludge particles with poor settling properties. The technique is employed at water purification stations and for fiber recovery on paper machines. It is also suitable for removal of suspended solids after biological treatment.

The sludge that accumulates on the surface has different properties from that settling to the bottom. As the surface sludge contains air, it cannot be pumped by centrifugal pump without the air first being removed. Air also has to be removed before different types of sludge can be mixed together, for example, during sludge handling. Flotation sludge has a solids content of 1%–3%.

Flotation clarification has these advantages:

- Small space requirement
- Effective for sludges with poor settleability.

Disadvantages include:

- High energy consumption
- Surface sludge difficult to mix with other sludges; high operating costs.

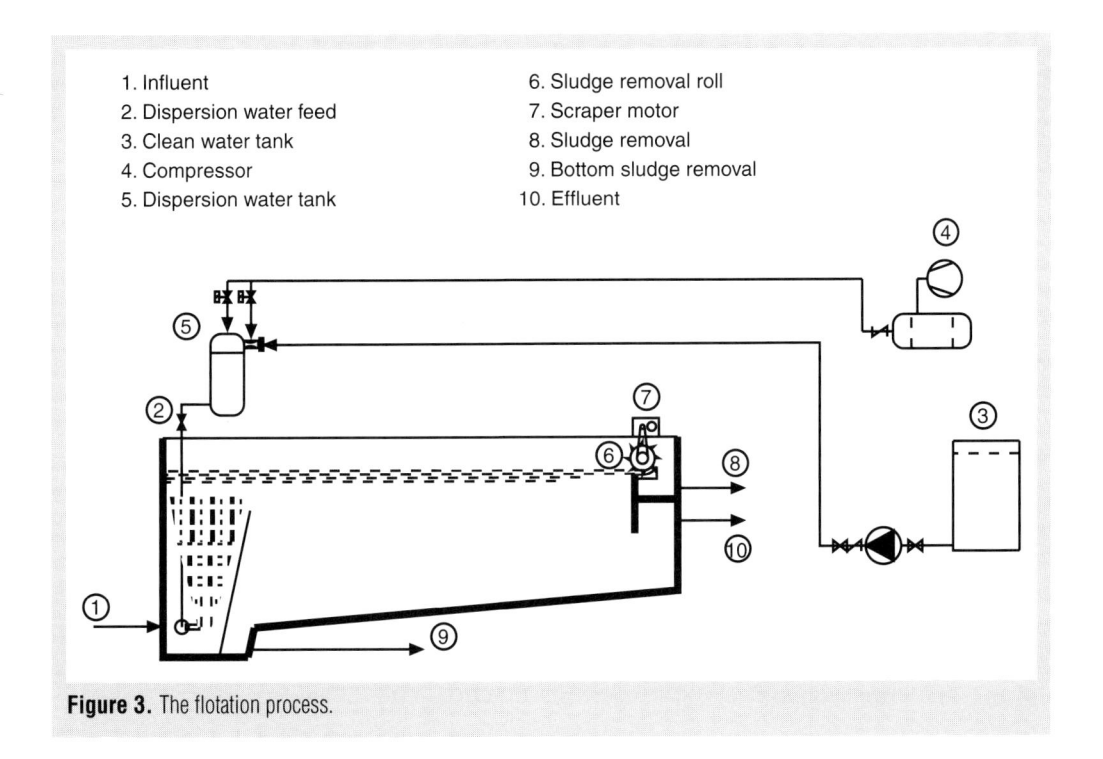

1. Influent
2. Dispersion water feed
3. Clean water tank
4. Compressor
5. Dispersion water tank
6. Sludge removal roll
7. Scraper motor
8. Sludge removal
9. Bottom sludge removal
10. Effluent

Figure 3. The flotation process.

CHAPTER 6

Flotation is employed for post-clarification at one activated sludge treatment plant in Finland. One fine paper mill uses flotation after chemical flocculation to clarify wastewater containing fiber and coating color particles.

The clarification tank used for flotation also is designed for particular surface loads, usually between 4 and 8 m^3/m^2h. Proper air/solid ratio is normally find out emprically.

Straining and filtration can be used in fiber recovery and in the external treatment of effluents with high solids contents. However, finely suspended solids and pitch often cause problems by blocking the filter. Also, the costs are high. Filtration can be used internally within the process to remove solids from certain effluent fractions, in which case both the water and the solids are returned to the process.

Ultrafiltration is a process that allows emulsified, suspended, colloidal, and high molecular mass material to be removed from a solution by means of a porous membrane. Solvent molecules and low molecular mass compounds pass through the membrane, causing the larger material to become more concentrated.

A pressure of 0.1–1 MPa is used to bring about membrane separation. Ultrafiltration produces two fractions: concentrate and permeate. The concentrate contains all material unable to pass through the membrane, while the permeate contains those molecules that have passed through.

Separation is achieved mainly on the basis of molecular size. The cut-off value of the membrane is the smallest molecular mass that is retained. Commercial ultrafiltration membranes are available with cut-off values between 1 000 and 10 000 daltons. The membranes used to treat effluents from pulp bleaching, for example, usually have a cut-off value of 6 000–8 000.

Ultrafiltration processes are either continuous or batch processes. The latter usually are based on the "open loop" principle, in which the concentrate is fed back into the system. This allows the concentrate to be brought up to the desired concentration.

With a continuous ("feed-bleed") system, concentrate is bled off at a particular rate or concentration has been reached. The composition and flow rate of the permeate also remain fairly constant.

In practice, several continuous ultrafiltration units are often linked together in series to form a cascade. With this arrangement, the concentration of material in the concentrate is raised stagewise to the desired value. A single system may contain 4–8 units.

The main uses for ultrafiltration in the forest industry are:

- Treatment of effluent from sulfate pulp bleaching
- Treatment of circulating water in mechanical pulp production
- Removal of resin from sulfite pulping effluents
- Recovery of lignosulphonates
- Recovery of latex at mills producing coated paper.

Effluent treatment

The purpose of treating effluent from the extraction (E) stage of pulp bleaching is to achieve maximum possible reductions in COD and AOX loads as well as color.

In Finland (1997), ultrafiltration is in use at two paper mills. Worldwide, there are 20 or so ultrafiltration plants, either pilot or full scale, many of them for treating effluents containing coating color. Numerous pilot and laboratory trials have been performed with different types of equipment and different effluents from paper mills. A lot of work has been done with the alkaline effluents from pulp bleaching.

The capacity of an ultrafiltration unit is often expressed as the flux rate, i.e., flow rate per unit area of membrane. The higher the average flux rate that can be achieved, the smaller the area of membrane needed in the process. The average flux, and therefore the area of membrane needed, is the main design parameter. Pilot trials should always be carried out to determine the flux and percentage reduction achievable with a particular effluent, and also to decide on the best membrane material for the job.

Process design requires the following information:

- Volume and composition of the effluent to be treated
- Desired cut-off value
- Final concentration of material in the concentrate
- Composition of the permeate.

The cut-off value for the process is determined either as the reduction in a particular parameter or in terms of retention. Within certain limits, the cut-off value achieved depends on the cut-off value and type of the membrane used. In terms of COD reduction, for example, the cut-off value also depends on the composition of the solution and the process conditions.

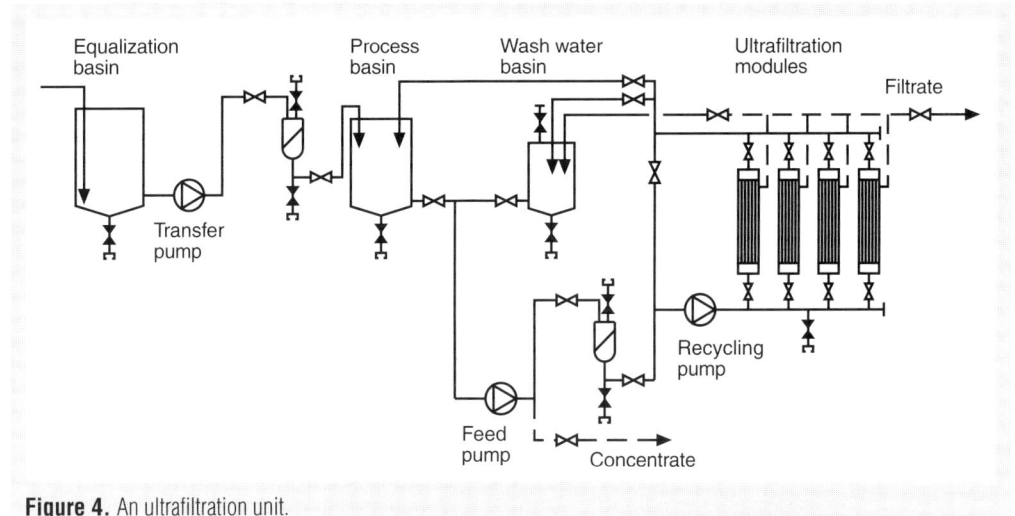

Figure 4. An ultrafiltration unit.

CHAPTER 6

The flux through the membrane diminishes as the solute content of the concentrate rises. Carrying out the process in several stages therefore improves the average flux rate and reduces the area of membrane needed. Raising the pressure across the membrane also increases the flux, but only up to a certain point. Flux rate also increases with increasing temperature. It may also be influenced by the pH of the solution.

There are several types of systems available like tubes and plates.

<u>Sand filtration</u> involves leading effluent through a 0.5–1 m thick bed of sand in order to remove solid particles from the water. The sand used has a particle size of 0.8–2 mm.

Traditionally, sand filtration is carried out in stages. The actual filtration stage takes from a few hours up to a day and is complete when the filter bed reaches a terminal headloss due to the buildup of material. The bed then has to be cleaned by backwashing, either at fixed intervals or whenever necessary. This stage takes about 15 minutes.

In cases where the effluent entering the filter contains relatively high amounts of solid matter, the filtration stage becomes shorter, and this reduces the performance of the filter.

In cases where the effluent volumes being treated are small or where contact with air needs to be avoided, pressure filters can be used. The principle is the same as with open filters, except that the entire system operates under pressure. In some filters, a portion of the sand can be regenerated continuously by washing.

The surface load used in open sand filtration of effluent is usually around 5 m/h. In the case of pressure filters, the surface load is somewhat higher.

Generally speaking, sand filtration is best suited for effluents with low solids contents. Carried out after biological treatment, sand filtration is capable of removing residual solid matter; hence, the organic matter and nutrients bound to the particles are also removed.

The advantages of traditional sand filtration are:

- Yields a filtrate with very low solids content
- Also removes COD, BOD, and nutrients bound to solid particles.

The disadvantages are:

- Process is unable to cope with effluents with high solids contents
- Washing the filter bed requires a large volume of water in a short space of time
- Filter needs maintenance to prevent slime formation.

Sand filtration is in some cases used for suspended solids and nutrient removal as a post treatment after biological treatment and flotation units.

<u>Continuous sand filtration</u> is designed so that filtration and washing of a portion of the filter bed are carried out simultaneously. Filtration itself takes place in the conventional way.

Some types of filter contain a number of small compartments separated by walls. The filter bed in each compartment can be flushed clean in sequence according to a pre-set program. In another type, sand is removed continuously from the bottom of the bed, taken through a separate washing section, and then returned to the filter.

In Finland, continuous sand filtration has been used to clean up raw water and, with varying degrees of success, to treat paper machine white water.

Flotation filters are a combination of a sand filter and a flotation clarifier in which flotation and filtration take place in the same tank. Flotation removes most of the influent solids, thus keeping the solids load to the filter bed fairly steady. Flotation filters are able to withstand prolonged surges in influent solids content.

The surface loads used in flotation filtration are similar to those in sand filtration. The process can be made more effective through the use of suitable chemicals. Flotation filtration is well suited for effluent whose solids contents vary a lot and also when the treated water has to meet strict quality requirements.

1.2 Chemical coagulation and flocculation

Chemical coagulation is combined with floc formation in a single process that involves rapid mixing in a flocculation tank having 1–3 compartments. The idea is to bring together small solid particles to form larger flocs, which are easier to remove. The addition of a polyelectrolyte during the rapid mixing stage assists flocculation. pH control is also important.

Vigorous mixing ensures that the chemicals required are mixed as completely as possible into the water before flocculation. The residence time of the effluent in the rapid mixing tank is 0.5–3 minutes.

Effluent in the flocculation tanks is kept moving by means of gentle stirring. Under the right conditions, solid particles coalesce to form larger particles, or flocs. Mixing must be performed gently because of the delicate nature of the flocs. The residence time of the water in the flocculation tanks is 20–40 minutes.

After chemical coagulation, the treated effluent is led into the clarification unit or other equipment designed for solids removal. Proper design of the system usually requires some prior experimentation.

The main chemicals used for coagulation are:

- Aluminium salts such as $Al_2(SO4)_3$ and $Al_n(OH)_mCl_{3n-m}$
- Iron (III) salts such as $FeCl_3$ and $Fe_2(SO_4)_3$
- Iron (II) salts such as $FeSO_4$
- Lime.

To achieve optimum flocculation results, it is often necessary to feed in a suitable polymer during the slow mixing stage.

The primary purpose of chemical coagulation is to neutralize the electrical charges on the particles and hence prevent repulsion. This generally means adding metal cations because the suspended organic material is almost always negatively charged. In most cases, the coagulant itself also precipitates out (for example, as the hydroxide), making the treatment more effective. For this reason, pH control is important. Organic polyelectrolytes, on the other hand, normally bind together dirt particles and in this way promote floc formation. Successful chemical coagulation requires knowledge of both surface chemistry and colloid chemistry.

The measurements most often used to determine the charge states are anionicity determination, zeta potential measurement, and calculation of particle sizes.

In the case of effluent treatment, chemical coagulation is applied mainly to the separate treatment of concentrated effluent fractions (e.g., those from bleaching, coating color preparation, and debarking). The flocs formed are removed from the water by sedimentation, flotation, or filtration.

Coagulation/flocculation is used mainly in the chemical treatment of raw process water. The advantages of chemical coagulation are its simplicity and low investment cost. The main drawback, however, is the large amount of sludge produced. As the sludge contains some of the chemical coagulants, it can seldom be returned to the process.

In Finland, chemical coagulation is used at a few paper and board mills in conjunction with gravitation settling and flotation treatment of fiber-containing process effluents. It also is used at some mills to treat effluents containing coating color.

2 Biological methods

Biological treatment is particularly useful for removal of low molecular mass organic matter. The process is based on the ability of microbes to live and reproduce in effluents. In so doing, microbes break down dissolved and colloidal substances by using them as nutrients. In this way, waste is converted partly into biomass and partly into carbon dioxide and water.

For biological treatment to succeed, it may be necessary to pre-treat the effluent. In particular, temperature, pH, and oxygen and nutrient contents must be suitable for microbial growth.

Biological effluent treatment processes can involve the simultaneous use of many different kinds of microbes ranging from the simplest bacteria to protozoa and even worms. Because of this, many processes are able to withstand disturbances and even toxic substances.

Quite a large number of forest industry plants use biological treatment (mostly the aerobic process).

2.1 Activated sludge process

In aerobic treatment, the breakdown of organic matter by microbes takes place in the presence of oxygen[2, 3]. The end products are carbon dioxide and water, and new cells. The biochemical reactions involved are complex, with enzymes playing an important role. In simplified form, the sequence of events is as shown in Fig. 5.

Effluent treatment

Figure 5. Principle of aerobic treatment.

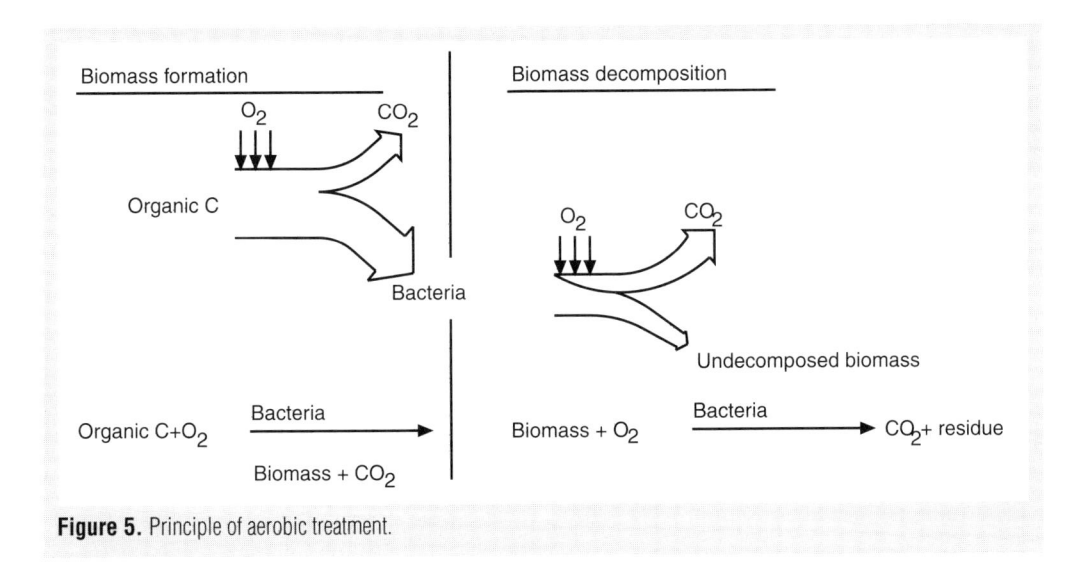

In the overall reaction, some 60%–70% of organic carbon is used to form new cells and 30%–40% is used to produce energy. One kilo of BOD in solution is sufficient nutrient to produce 0.4–0.7 kg of new biomass, a process that consumes 0.5–0.9 kg of oxygen. During the breakdown of the biomass, a process that also requires oxygen, some 20% remains intact.

The key factors for the success of an aerobic process are:

- Amount of nutrients in relation to amount of biomass
- Type of nutrients (sorption by biomass)
- Temperature, pH, oxygen content, and concentration of toxic substances.

The nature of the effluent, i.e., the types of nutrient present, is reflected in the sorption onto the biomass and in the rate of decomposition.

Raising the temperature greatly speeds up the rate of reaction. Decomposition is often 3–5 times faster at 35°C than at 20°C.

The best pH for this process is 7–7.5, while the range 6.8–8 is workable. A low pH favors the growth of yeasts present in the biomass. A properly working biological process has considerable buffering capacity due to the formation of carbon dioxide.

A sufficient nutrient to organic carbon ratio is also important. The old basic rule based on the amount of biomass produced is BOD:N:P =100:5:1. Certain trace elements (e.g., metals like iron, potassium and zin) are also needed, but effluents usually contain sufficient levels of these.

In the aerobic process, the water or water-sludge suspension in the reactor is supplied with an excess of oxygen by means of aeration.

CHAPTER 6

The main aerobic effluent treatment methods used are:

I. Activated sludge processes

- Continuous (plug flow or complete mixing, see Eckenfelder et al[2])
- Batch process

II. Lagoons

- Aerated
- Stabilization ponds

III. Fixed bed processes

- Biofilter (fixed or mobile media)
- Biorotor

IV. Nitrogen and phosphorus removal processes.

The processes used most frequently by the Finnish forest industry are activated sludge treatment (plug flow) and aerated lagoons (also plug flow). The processes have been chosen because they are effective with forest industry effluents.

In activated sludge treatment, the biomass (i.e., activated sludge) and the effluent are mixed to give an almost homogeneous suspension in the reactor, into which air is led (see Fig. 6).

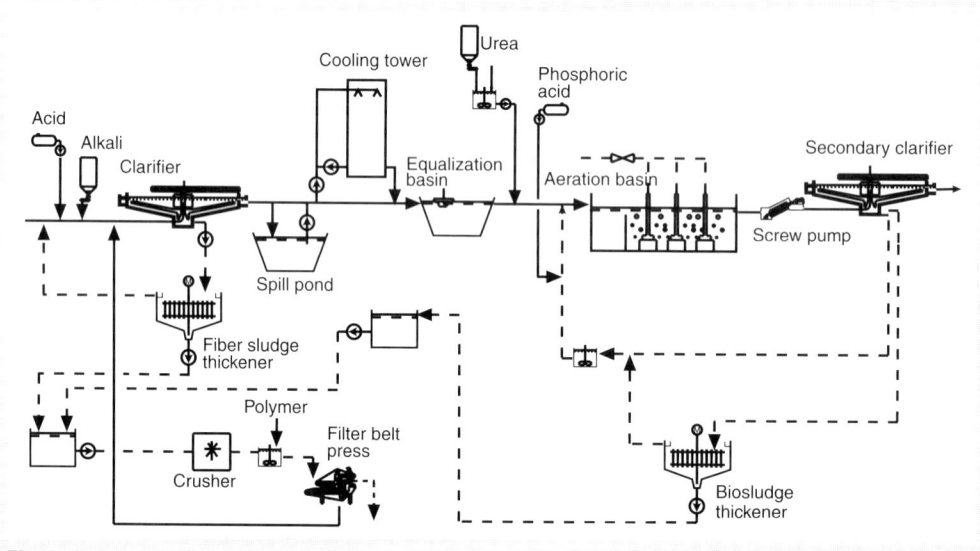

Figure 6. Principle of the activated sludge process.

The sludge-water suspension is led out of the reactor (aeration tank) to a clarifier, where the sludge settles to the bottom. Most of this sludge is pumped back into the aeration tank. The excess is removed from the process, and this provides an important way of controlling the process.

The activated sludge process has two main components: aeration and sedimentation. How these function in unison determines the effectiveness of the entire process.

The following concepts are used in dimensioning an activated sludge plant:
Sludge load refers to the ratio of BOD to the amount of biomass in the aeration tank, i.e.,

$$F/M = (KgBOD_7)/d/(kgMLVSS) \tag{3}$$

where $MLVSS$ (mixed liquor volatile suspended solids) is the amount of active or organic solids in the aeration tank. This concept is used in order to take into account the inert sludge, which is present but does not take part in the process. However, the use of $MLVSS$ has not been very successful in the case of forest industry effluents, as biologically inert material such as wood fiber is included in $MLVSS$.

Sludge load is often expressed in relation to total solids load, which then includes inorganic and other nonactive solids. Sludge load is crucial to the dimensioning of the process as it determines the capacity needed for the aeration tank. Under normal loadings, sludge loads of between 0.1 and 0.4 kg BOD/kg sludge obtain the best operation.

There is also a correlation between sludge load and sludge retention time SRT used for controlling activated sludge plants.

Sludge content is the dry weight of solids per unit volume of sludge in the aeration tank. Normal values are 2.5–4 kg/m^3.

Sludge index (sludge volume index, SVI, or preferably diluted sludge volume index, DSVI, units ml/g). This reflects the settling properties of activated sludge. The target value is usually around 120. If SVI exceeds 200, we speak of bulking.

Floc load (units g COD$_{in}$/g of activated sludge) is a measure of how much of the incoming COD load is sorbed by the biosludge[2].

Oxygen content (mg O_2/l) in the aeration tank should be above a certain minimum level (1.5–2.0 mg/l in activated sludge plants using air) in order for the process to function properly. The oxygen content often varies at different points in the aeration tank, with the extent of the variation depending on the dimensions of the tank.

Circulation number (CN) indicates the average number of times during its lifetime biosludge is circulated through the secondary clarifier. An approximate equation for calculating CN is $Qr/Qw + 1$, where Qr is the quantity of return sludge and Qw is the quantity of excess sludge wasted from the system.

Sludge age (SRT) is the ratio of the amount of solids in the aeration tank to the amount of solids wasted from the system. Sludge age is normally 5–15 days .

Volumetric loading indicates the amount of biodegradable organic matter entering the aeration tank per unit volume of influent. Volumetric loading is the product of sludge load and sludge content of the influent.

CHAPTER 6

Surface load to clarifier
The clarifier is a vital part of the activated sludge process. The following parameters are used in conventional dimensioning:

- Hydraulic surface load
- Surface sludge load
- Volumetric sludge load.

The hydraulic surface load is

$$S_H = Q/A \tag{4}$$

where S_H is Surface load, m³/m²h (or m/h)
  Q Flow rate, m³/h
  A Surface area of clarifier, m².

The values normally recommended to be given omit the contribution from the return sludge flow. The appropriate value depends on the sludge loading to the aeration tank and the shape of the clarifier.
The surface sludge load :

$$S_L = M/A \tag{5}$$

where M is Amount of activated sludge entering the clarifier per hour.

The volumetric sludge load

$$S_{SVI} = S_H \times C \times SVI \times 10^3 \tag{6}$$

where S_H is Hydraulic surface load
  C Concentration of activated sludge
  SVI Sludge volume index
  S_{SVI} Volumetric sludge load.

However, recent measurements and other investigations have shown that the surface load concept is inadequate for proper dimensioning of clarifiers. Effluent retention time is obviously a useful way of evaluating the performance of a clarifier.

Sludge return ratio
Return ratio is a concept that the forest industry has adopted from the treatment of municipal wastewaters. The ratio depends on:

- Consistency of return sludge
- Target for sludge content in aeration section.

The consistency of the sludge being returned to the process depends on the sludge's settling properties, which in turn are determined by the type of effluent being treated and the design of the activated sludge process.

A concept that should be used alongside, or preferably in place of, return ratio is floc load. However, some useful data expressed in terms of return ratio can be found in the literature.

Excess sludge

The sludge in an activated sludge process consists of biomass and solids entering the process. This mixture is referred to as biosludge. The growth of biomass in an aerobic process can be calculated from the following empirical equation:

$$dS = aSr - bX \qquad (7)$$

where
- dS is Formation of biomass, kg (VSS/d)
- Sr BOD_7 reduction (kg/d)
- X Amount of suspended solids in aeration tank (kg VSS)
- a Coefficient for formation of solid material
- b Rate constant for endogenic decomposition.

According to the equation, the formation of new biomass is related to the BOD reduction, the amount of solids in the aeration section, and the rates of formation and decomposition of solid material. In calculating a material balance, account naturally has to be taken of the mass flows entering and leaving with the effluent.

Modifications of the activated sludge process

The classification of processes according to loading is similar to that used in municipal wastewater treatment, and is as follows:

- Extended aeration, $F/M = 0.05–0.1$
- Low-load process, $F/M = 0.1–0.3$
- Normal-load process, $F/M = 0.25–0.4$
- High-load process, $F/M > 0.6$.

The extended aeration plant is suitable for treating forest industry effluents. The plant has good tolerance to shock loads. There is also normally sufficient time for making adjustment during operating. If land basins can be used for aeration, as this makes the costs reasonable compared to other activated sludge processes.

The activated sludge plants used in the forest industry are of the low or normal-load type. Treatment is complete (effluent dissolved BOD < 10 mg/l) because there is an adequate amount of biomass and the retention time is also adequate.

CHAPTER 6

Investment cost for high-load plants are generally lowest. Operating is not normally easy and biosludge is formed in large amounts. The use of high-load plants has greatly decreased e.g. in Finlad during the last ten years .

Classification according to aeration:
- Plug flow
- Complete mixing
- Biosorption (contact stabilization), also known as sludge aeration process
- Tapered aeration
- Step aeration
- High-purity oxygen processes (or partial oxygen process)

In a plug flow arrangement, effluent and return sludge are fed into the aeration tank. The oxygen demand is highest near the inlet and decreases steadily as the mixture flows through the tank. The effluent/sludge mixture passes along the long tank as a "plug."

In a complete mixing system, the idea is to mix the incoming effluent as completely as possible in all parts of the aeration tank. This means the oxygen demand is the same throughout the tank. A peak loading is distributed evenly throughout the tank. The biological part of the process is easy to control.

In the contact stabilization process, return sludge from the settling tank is first pumped to a re-aeration tank, where the micro-organisms metabolize the nutrients extracted from the waste. Contact stabilization has been used where there are large variations in effluent flow or BOD loading. However, in the light of present knowledge and experiences e.g. from Finland the use of this process for pulp and paper mill effluents should be regarded with some skepticism. More basic studies will be needed.

In the tapered aeration process, the idea is to supply oxygen according to demand so that aeration is more vigorous near the inlet. In a normal aeration tank, the oxygen demand is about three times greater at the inlet than at the end of the tank, as biomass growth is fastest near the inlet.

To avoid a shortage of oxygen near the inlet, effluent can be fed into the tank at several points along its length. The BOD loading is in contact with the entire amount of sludge, as in the case of complete mixing. Effluent can be excluded from the initial part of the tank, which can then be used for sludge aeration as in the contact stabilization process. One practical advantage of the step aeration arrangement is that the detrimental effects of poor diffuser positioning can be greatly reduced.

Oxygen plants do not differ in principle from conventional activated sludge plants. The aeration section is kept under slight pressure by adjusting the feed of pure oxygen. The first such plant was built in the United States in 1975.

Effluent treatment

Two-stage activated sludge treatment processes have been in use for about 30 years. The first stage involves a low oxygen content combined with a high sludge loading. The second stage is then a conventional activated sludge process designed according to the type of effluent being treated. The advantages of a two-stage process are claimed to be its tolerance to variations in loading and effluent toxicity, and the smaller aeration tanks required.

Since the late 1980s, aerobic selectors have been used with success to improve the removal of organic matter. By splitting the aeration tank into small compartments, sorption of organic matter by biomass is more effective than in the case of a normal plug flow type of process. The aerobicity of the selectors has been found to be of vital importance in the case of forest industry effluents[4].

a) Staggered aeration

b) Staggered effluent feed

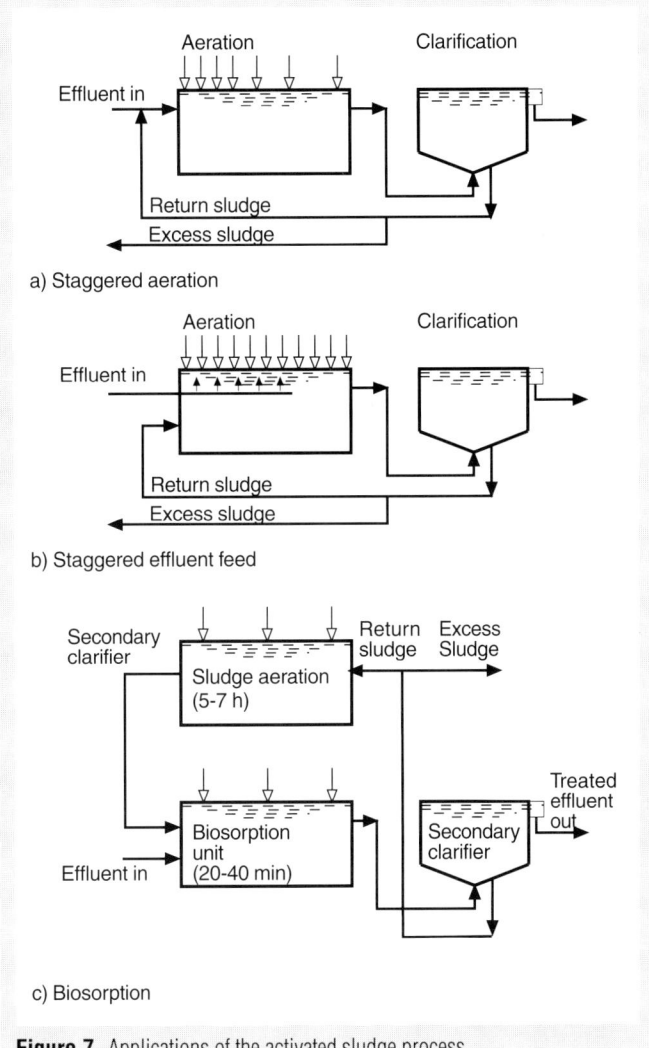

c) Biosorption

Figure 7. Applications of the activated sludge process.

Activated sludge processes are influenced by a number of factors as follows:

Oxygen
The activated sludge process is designed to operate under aerobic conditions.

Temperature
The rate of a biochemical reaction depends on temperature. This has been expressed using the following equation:

$$k_t = k_{20} \times Q^{t-20} \tag{8}$$

where k_t is Rate constant for reaction at temperature t
        k_{20} Rate constant at 20°C
        Q Temperature coefficient.

CHAPTER 6

The values obtained for the temperature coefficient generally lie between 1.0 and 1.03.

The activated sludge process functions well over a wide temperature range. However, rapid temperature changes cause problems. Used to treat municipal wastewaters, the activated sludge process has operated at temperatures as low as 2°C–5°C. Effluents from the wood- processing industry are usually warm, and the activated sludge process still performs well at up to 35°C–37°C. Above 40°C, however, the effluent first has to be cooled.

Nutrients

The concentration of nutrients in relation to that of organic matter is important in effluent treatment. Effluents from the wood-processing industry generally have a BOD:N:P ratio of 100:(1–2):(0.15–0.3) (see chapter 4,table 1) . Biological treatment requires certain minimum concentrations of nutrients (mainly phosphorus and nitrogen), and these often have to be added. The activated sludge process, for example, requires adding 0.8%–3.5% of nitrogen and 0.3%–0.6% of phosphorus based on effluent BOD.

Probably the most reliable way of calculating the amounts of nutrient that have to be added is via a nutrient balance (nutrients in equals nutrients out). For calculation purposes, the phosphorus content of the treated effluent is taken to be 0.1–0.2 mg/l and the nitrogen content is 2–2.5 mg/l[4].

Insufficient nitrogen and phosphorus levels result in sludge with poor settling properties, lower BOD and COD reductions, and impaired sludge formation. Addition of excess nutrients, on the other hand, increases the loading to the recipient.

Toxic substances

Inhibition of the process due to toxicity may depend either partly or completely on the concentration of toxic substances. The effects these toxic substances have on the system depend on factors such as temperature, nutrient level, types of micro-organism present, and the ability of the micro-organisms to adapt. The hydrogen sulfide in sulfate pulp mill process discharges and the sulfur dioxide from pulp bleaching are highly toxic. Fatty acids and resin acids are also toxic to activated sludge.

Process discharges from the wood-processing industry do not normally give rise to toxicity problems, although occasional high discharges of organic compounds from mills can disrupt the treatment process.

Color usually is not reduced by activated sludge treatment. However, in certain cases reductions of 20%–50% are possible.

Aerobic processes can yield reductions of 20%–60% in chemical oxygen demand, depending on the nature of the effluent, whereas the target for biochemical oxygen demand is a reduction of 85%–95%. Reductions of over 95% generally require multistage treatment.

There are several ways to calculate the necessary dimensions of an activated sludge treatment plant; one of these (Eckenfelder et al.) is described below.

Aeration tank capacity. The units used have been converted to the metric system. It is

Effluent treatment

assumed that the plant being designed is of the complete mixing type. The calculation is based on the equation for the decomposition of organic matter:

$$\frac{S_0(S_0 - S_e)}{X_v t} = K(S_e - y) \tag{9}$$

where S_0 is Total BOD, COD or TOC of influent, mg/l
S_e Soluble BOD, COD or TOC of water out, mg/l
X_v Volatile suspended solids (MLVSS) concentration, mg/l
t Time in aeration tank
K Rate coefficient for breakdown of organic matter, d^{-1}
y Undecomposed portion of organic matter, mg/l.

When BOD is used as a measure of organic content, y is usually taken as zero. The rate coefficient, K, depends on temperature as follows:

$$K_2 = K_1 \Theta^{(T_2 - T_1)} \tag{10}$$

where K_1 and K_2 are the biodegradation rate coefficients at temperatures T_1 and T_2, respectively
Θ is Temperature coefficient.

Θ normally lies between 1.03 and 1.09, but the practice is to determine its value empirically.

Tests are also needed to determine the sludge load, F/M, and sludge age, G, on which the capacity of the aeration tank depends. Sludge load, F/M, is determined as follows:

$$F/M = S_0/X_v t \tag{11}$$

and sludge age, G as follows:

$$G = \frac{X_v}{\Delta X_v} \tag{12}$$

where S_0 is Total BOD, COD or TOC of influent, kg/d
X_V Average MLVSS in the aeration tank, kg/d
ΔX_V Production of excess biosludge, VSS, kg VSS/d
t Time in aeration tank, d.

Once the necessary sludge load has been determined, the aeration time and aeration tank volume can be obtained from Eq. 11. The longest aeration time required is

CHAPTER 6

also checked using Eq. 9, and the highest of the values thus obtained is used in the plant design.

Note that the necessary sludge load is determined with a view to obtaining sludge with optimum settling properties.

Biodegradability of biosludge. Laboratory tests must be conducted at the design stage to determine what proportion of the biosludge can be broken down biologically. Adams et al. have reported a figure of 77%. The proportion, X, of the organic matter that does not biodegrade increases along with sludge age. This proportion can be obtained from the following equation:

$$x = \frac{aS_r + bX_v - \sqrt{(aS_r + bX_v)^2(0.77aS_r)}}{2bX_v} \tag{13}$$

where x is Biodegradable portion of MLVSS
  a Sludge yield coefficient, kg VSS produced/kg organic matter removed
  S_r Organic matter removed (BOD, COD, TOC), kg/d
  b Sludge decomposition coefficient, kg VSS decomposed/kg MLVSS in the aeration tank x d^{-1}
  X_v Average MLVSS in aeration tank, kg

x is needed in calculating the nutrient and oxygen requirements.

Nutrient requirement. The biomass formed through the decomposition of organic material contains 12.3% nitrogen and 2.6% phosphorus. As sludge age increases and the biomass decomposes, its nitrogen content falls to 7% and its phosphorus content to 1%. These values can be used to calculate the nitrogen and phosphorus requirements as follows:

$$\text{Nitrogen requirement (kg/d)} = \frac{0.123 \times \Delta X_v}{0.77} + \frac{0.07(0.77-x)\Delta X_v}{0.77} \tag{14}$$

$$\text{Phosphorus requirement (kg/d)} = \frac{0.026 \times \Delta X_v}{0.77} + \frac{0.01(0.77-x)\Delta X_v}{0.77} \tag{15}$$

where ΔX_v is production of excess sludge, kg VSS/d

Oxygen requirement. The following four types of consumption determine oxygen requirement:

a) Oxygen needed for biodegradation of organic matter ($a'S_r$)

b) Oxygen consumed in cell breakdown ($b'X_v$)

c) Immediate chemical oxygen demand (R_c)

d) Oxygen needed to oxidize nitrogen compounds to nitrate (R_n).

Effluent treatment

The total oxygen requirement is thus

$$R_r = a'S_r + b'X_v + R_c + R_n \qquad (16)$$

where
- R_r is Total oxygen requirement, kg O_2/d
- a' Oxygen consumed in cell synthesis, kg O_2/kg of organic matter removed per day
- b' Oxygen consumed in cell breakdown, kg O_2/kg MLVSSd
- R_c Immediate chemical oxygen demand (measured), kg O_2/d
- R_n Oxygen consumed in oxidation of nitrogen compounds, kg O_2/d
- X_v Average MLVSS in aeration tank/d.

Biosludge production. Some of the biosludge produced in the process has to be removed from time to time for drying and disposal. In calculating the rate of removal, the following must be considered:

a) Non-degradable suspended solids entering the activated sludge plant, SS (fX_i)

b) Growth of biosludge through cell synthesis ($a\,S_r$)

c) Breakdown of biosludge due to the decomposition of biosludge (bX_v)

d) Sludge escaping with the treated effluent (X_e).

Production of excess sludge is obtained from the equation:

$$\Delta X_v = aS_r + b_x X_v \qquad (17)$$

and total sludge production from the equation 18.

$$\Delta X = fX_i + \frac{\Delta X_v}{f_v} - X_e \qquad (18)$$

where
- ΔX is Sludge production, kg SS/d
- f Proportion of nonbiodegradable solids in influent
- X_i Suspended solids entering with influent, kg SS/d
- f_v MLVSS in relation to $MLSS$ in aeration tank, $MLVSS/MLSS$
- X_e Suspended solids in effluent out, kg SS/d
- ΔX_v Production of excess biosludge, kg VSS/d
- X_v Average MLVSS in aeration tank.

Empirically derived formulae are also used to calculate the temperature in the aeration tank in winter and summer.

Calculations such as these, although based on sound assumptions, require that the coefficients used in the different formulae be determined experimentally in the laboratory at the design stage of the plant. Information on the operation of existing plants can also contribute to the determination of the coefficients. Existing plants also afford an opportunity for the causes of any disturbances to be investigated.

CHAPTER 6

Disturbances may be caused by poor process control, the treatment of effluent with unusual characteristics, or by deficiencies (either temporary or continuous) in the operation of equipment.

The first step in activated sludge treatment is sorption, in which solids, colloidal substances, and dissolved compounds become bound to the biosludge. Return sludge from the secondary clarifier settles well and its growth is encouraged. The sludge return rate must be adequate in relation to the rate of entry of sorbable material in the influent. This is usually calculated as kg COD_{inf}/kg return sludge per unit time. The terms floc load[4] and impact load are also used. Sorption of material by return sludge is best accomplished in selectors or in plug-flow tanks. A common control error in the case of concentrated effluents from the forest industry is to **exceed the floc load**. Organic material is sorbed later in the process, even by sludge with poor settling properties.

Forest industry effluents contain too little nitrogen or phosphorus, or both, to meet the nutrient requirement in the treatment process. The **incorrect dosing of nutrient** is a common cause of problems in the activated sludge process (although the situation is improving). This affects sorption in the succeeding biodegradation process.

In plug-flow plants, oxygen consumption generally reaches a maximum 1.5–3 hours after sorption has taken place, falling to its minimum some 6–8 hours after sorption. This demands the correct distribution of aeration capacity and, of course, adequate mixing. About 50% of the aeration capacity generally has to be provided in the first third of the tank. Nevertheless, it is often the case that **aeration capacity is not properly allocated** along the tank. To overcome this, effluent is sometimes fed into the aeration system stepwise to even out the oxygen loading along the tank. However, this carries the risk of sorption of material by biomass with poor settling properties.

The plant's **aeration capacity can be exceeded if the sludge is retained in the system** too long, i.e., the amount of sludge requiring oxygen becomes too great. Raising sludge age reduces biomass production, increases treatment efficiency, reduces the need for added nutrients, and makes correct dosing less critical. Raising the sludge age is thus beneficial, provided the aeration capacity is not exceeded.

Other, less common, causes of malfunctioning of the activated sludge process include **incorrect pH control, high concentrations of resin acids and fatty acids, and shortages of certain trace metals.**

Of the toxic chemicals used in manufacturing, the **peroxide and sulfur dioxide water** employed in pulp bleaching, especially when used in large amounts, have also caused disruptions to activated sludge treatment in some cases. At certain plants where a lagoon or biofilter is linked in series with an activated sludge plant, the treatment efficiency of the activated sludge process has been found to be below normal. The main problem is the abnormally high suspended solids leaving with the treated effluent. This could be caused by fairly high levels of poor-settling biomass entering the plant.

Tests have shown that of the metals entering the process, sorption of iron, zinc, manganese, and potassium is particularly effective. It has also been found that addition of metals such as iron can improve the plant's performance.

In terms of hardware, most critical for successful treatment are primary sedimentation, the aerators and their location, a secure power supply to the compressors, provi-

Effluent treatment

sion of a standby compressor, and sludge drying. Inadequate or malfunctioning on-line measuring instruments are often the reason for incorrect process control. Most commonly this applies to measurement of flow rate and oxygen level. There also may be serious shortcomings in relation to the dimensioning of aeration tanks and secondary clarifiers.

2.2 Other aerobic treatments

Aerated lagoons have been used for the treatment of warm industrial effluents. An aerated lagoon is similar to an activated sludge process without sludge recycling. The BOD reduction obtained depends on the type of effluent (rate of decomposition) and retention in the system.

The biomass produced in an aerated lagoon has poor settling properties. Some can be removed by settling or by flotation.

Aerated lagoons are cheaper to construct and operate than activated sludge plants provided that sufficient space is available and the ground is suitable (e.g., a dam can be built against the recipient watercourse). The BOD reduction obtained in northern Europe is up to 90% (in winter 75%–80%).

1. INCOMING EFFLUENT
2. RECYCLING BASIN
3. PUMP
4. FILTER TOWER
5. DISTRIBUTOR
6. AIR INTAKE SLOTS
7. COLLECTION BASIN AND CHANNELS
8. CLARIFICATION BASIN
9. EXCESS SLUDGE
10. EFFLUENT TO RECEIVING WATER OR NEXT TREATMENT STAGE

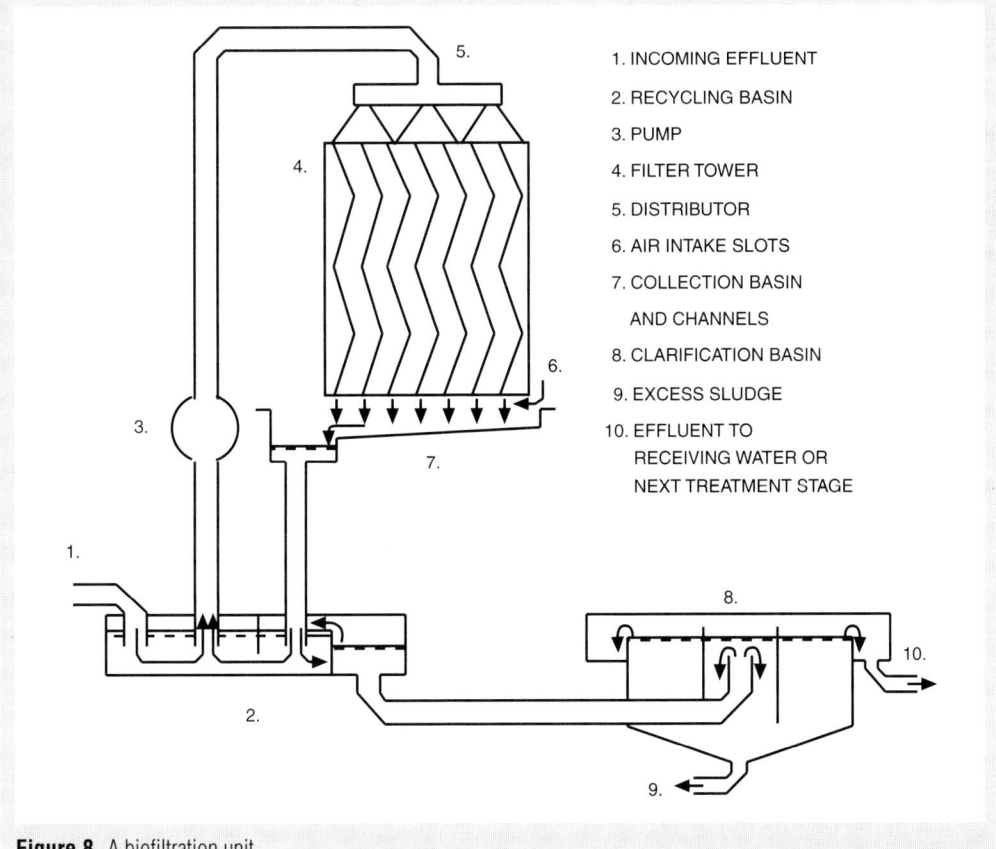

Figure 8. A biofiltration unit.

CHAPTER 6

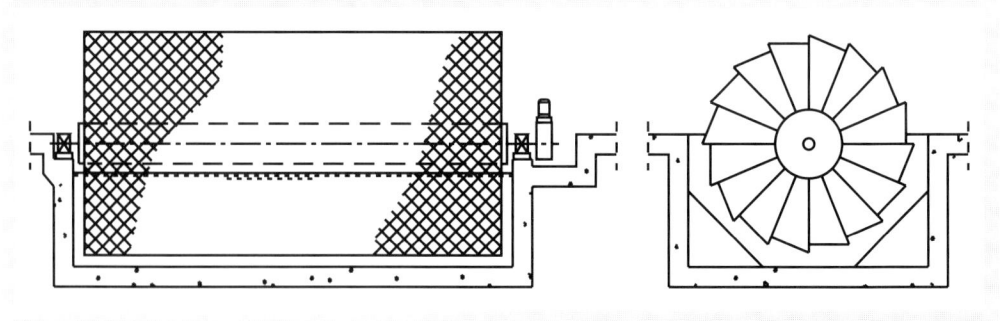

Figure 9. A biorotor.

In processes employing a solid biofilm, the biomass forms a layer on the surface of a solid support medium (thin-layer fermentation). Effluent is made to trickle across the surface of the support medium (biological filtration, biorotor) or else the medium moves through the effluent (biorotor, methods based on fluidization). The design of a biofilter is shown schematically in Fig. 8 and of a biorotor in Fig. 9^3.

Growth of biomass causes the film to become thicker, resulting in anaerobic conditions in the layer attached to the medium. This causes detachment of the film of biomass, which is removed in a clarifier as excess sludge.

In biological filtration, effluent is pumped onto a bed of inert granular material, over which it flows in the form of a film. Air passes upward through the filter tower by normal ventilation, i.e., counter-current to the effluent flow. Effluent leaving the tower goes to a clarifier. Recycling and clarification can be arranged in several ways.

At one time, gravel was used as the inert filter medium. Today, the use of a plastic matrix or plastic granule system is preferred. The idea is to achieve a high contact area (100 or even 200 m^2/m^3).

Biological filter capacity is determined by BOD load, but also by the return ratio.

The main advantages of biofiltration over activated sludge treatment are that the process is immune from bulking problems and that its energy costs are only one-third to one-quarter of those of an activated sludge plant of the same capacity. Also, a biofilter is largely unaffected by cold temperatures, provided it is properly designed. On the other hand, problems have been encountered with blockages, even with plastic media.

Biofiltration could be a viable option in the following cases:

- As a pre-treatment for concentrated effluent (600–700 mg BOD/l)
- For partial biological treatment (reduction of 50–80%)
- For pre-treatment of warm effluent (biofilter than also acts as a cooling tower).

A biorotor normally used for small or for medium size plants consists of a fixed medium in cylindrical form rotating on a horizontal shaft in a tank. It is slowly rotated so that either 40% or 60% of its surface is submerged. The bio-film formed on the medium is thus alternately in contact with effluent and then with air. The shear forces acting on

Effluent treatment

the film cause some of the film to be removed. There are usually 3–5 rotors or cascades connected in series.

2.3 Anaerobic processes

General

In anaerobic processes, organic matter is broken down by microbial activity in several stages in the absence of oxygen. The substrates are organic compounds and oxygen-containing compounds. Anaerobic activity leads to the formation of new cells and the release of energy for use by the cells. The products are methane, carbon dioxide, water, and other gases such as hydrogen, nitrogen, and hydrogen sulfide (Fig. 10). An aerobic treatment system is shown in Fig. 11[3].

Anaerobic processes operate in two temperature ranges: the mesophilic range (29°C–38°C) and the thermophilic range (49°C–57°C). Although reaction rates are faster in the thermophilic range, the susceptibility to disruption makes this the less preferred choice.

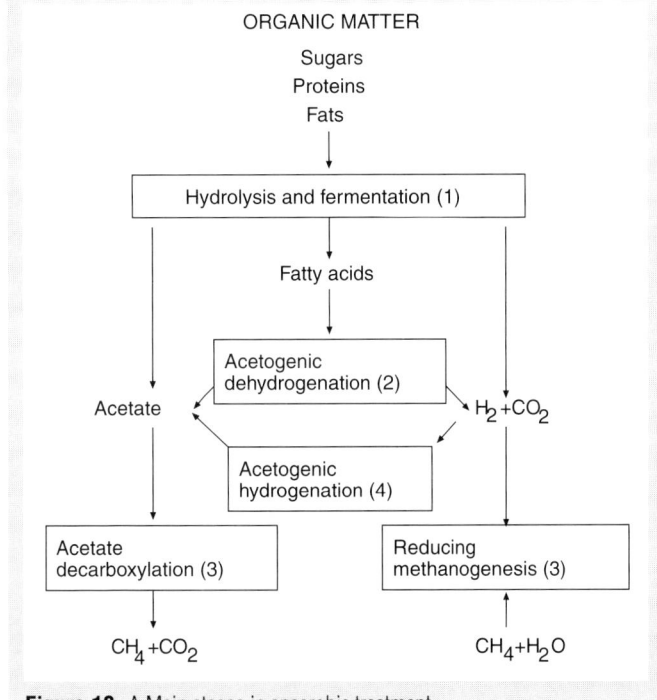

Figure 10. A Main stages in anaerobic treatment.

Anaerobic treatment takes place in three main stages:

1. Hydrolysis/liquefaction

2. Formation of acids

3. Methane formation.

Methane formation takes place in the pH range 6.6–7.6, the optimum being around 7.0. Outside this range, methane ceases to be generated and the carbon dioxide content of the system builds up. Thus, pH control is vital. The agents most often used for pH control are lime and sodium bicarbonate. Several inorganic salts, when present at low concentrations, have been found to stimulate the process.

CHAPTER 6

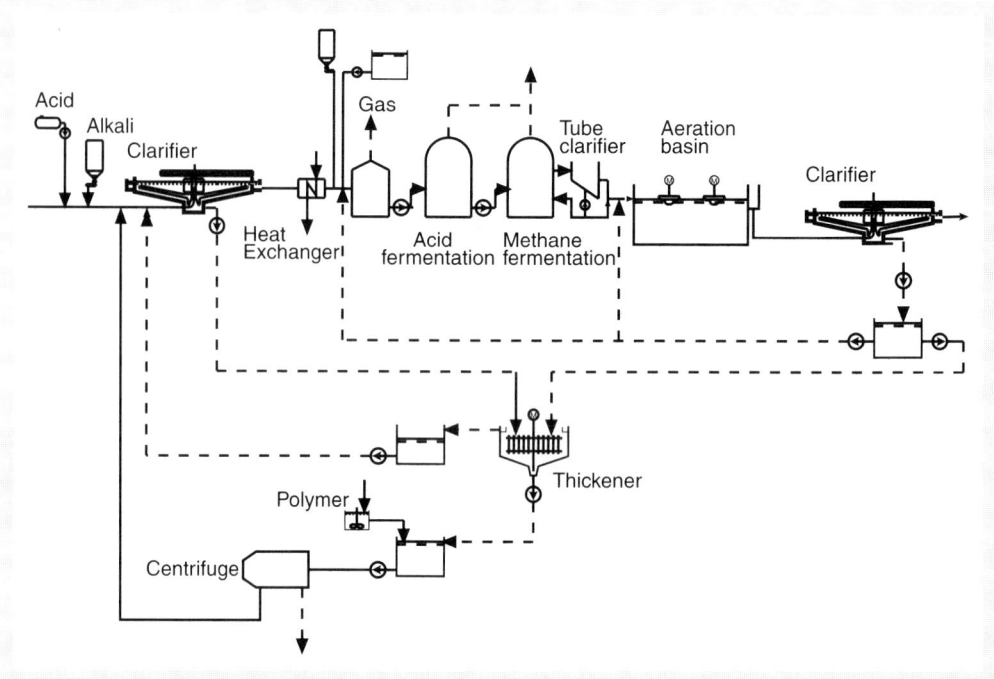

Figure 11. An anaerobic treatment plant.

Anaerobic processes have long been used in the food and brewing industries. They have also been used for the treatment of sludges, and more recently have been applied to forest industry effluents. The use of anaerobic treatment for the further breakdown of biosludge from an activated sludge plant has not yielded encouraging results. However, certain forest industry effluents such as condensates have been successfully treated.

There are two main types of anaerobic reactor:

I. Those in which the microbes are mixed uniformly with the effluent (sludge reactor)

II. Those in which the microbes are retained on a medium of some kind (biofilm reactor).

The first and simplest reactor, or fermenter, to be introduced was a tank, either with or without heating. There is generally a long hydraulic retention time to ensure that solid organic matter has time to decompose. The lengthy biological treatment of effluent was originally anaerobic treatment in a septic tank. The BOD_7 reduction achieved with this method is 60%–80%, depending on the treatment time. The hydraulic retention times used are several days or more.

Effluent treatment

The high-rate process is used particularly for anaerobic stabilization of agricultural sludges. The reactor resembles a septic tank, with the difference that the slurry is mixed. It is used for stabilization of sludges from municipal treatment plants.

Further improvements have resulted in the contact process (Fig. 12). The principle resembles activated sludge treatment under anaerobic conditions followed by clarification. This process is suitable for effluents with high suspended solids contents and for very concentrated effluents.

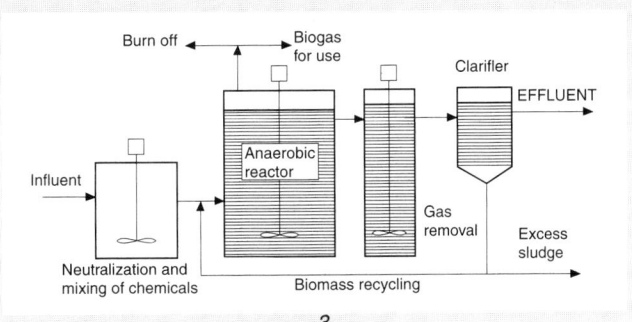

Figure 12. Anaerobic contact method[3].

Anaerobic filtration (Fig. 13) has undergone rapid development in recent years. The packed filter medium may be either a matrix, as in the aerobic filter, or it may be composed of discrete particles. In the case of concentrated effluents, recycling may be necessary; whereas, with dilute effluents, a single pass through the filter is enough.

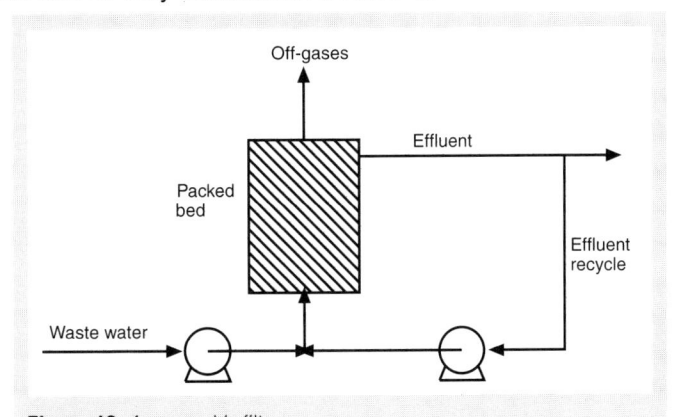

Figure 13. An anaerobic filter.

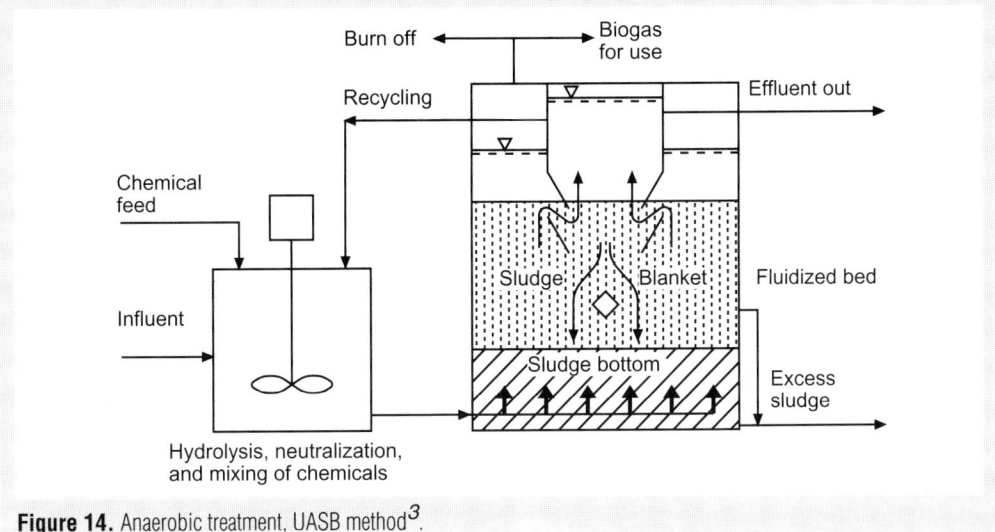

Figure 14. Anaerobic treatment, UASB method[3].

CHAPTER 6

The main reactor types used by the forest industry are anaerobic fermentation using a fluidized bed (Fig. 14) and the upflow anaerobic sludge blanket (UASB), in which the blanket is composed of granules. Both types have yielded good results.

Advantages of anaerobic treatment:

1. No aeration system is needed, and this cuts investment and operating costs.
2. Less cell mass is produced than in aerobic processes.
3. The process is suitable for several different types of waste.
4. There is a low nutrient requirement.
5. Oxygen is supplied by sulfates and nitrates.
6. The mixture of methane and carbon dioxide produced can be burned for energy.

Disadvantages of anaerobic treatment:

1. Higher than ambient temperatures are needed.
2. The short hydraulic retention time necessitates a high biomass concentration.
3. Production of sufficient methanogenic bacteria takes 2–11 days, which means high retention times are needed for solids.
4. The process is sensitive to sudden changes in conditions.
5. The process needs an induction period of 20–30 days before equilibrium is reached.
6. There are odor problems.
7. Final handling of treated effluent is difficult (odor, BOD).
8. Anaerobic processes always need to be operated in conjunction with an aerobic process.

Anaerobic prosesses have been succesfully used for printing paper and board mill effluents.

3 Other treatment methods

3.1 Activated carbon adsorption

Activated carbon has long been used in cleanup processes because of its ability to adsorb certain compounds from solution. This is attributed to the high specific surface area (600–1000 m^2/g) and porous structure of carbon.

The main types of activated carbon adsorption process are:

- Adsorption onto granular carbon in a column
- Adsorption onto powdered carbon in a tank followed by clarification (PAC).

Effluent treatment

A continuous adsorption process on a column is fairly simple to arrange. The organic components responsible for color are gradually adsorbed into the pores of the activated carbon. Once saturated, the carbon has to be regenerated before re-use.

The saturated carbon is removed from the column, dried, and heated. The carbon regenerated in this way is then returned to the column. The regeneration losses on heating are 5%–15%.

Carbon has a certain capacity to remove organic material. The carbon requirement is usually 3–10 kg/m^3 of effluent although more is needed as the contact time decreases.

Tests with activated carbon have been conducted for the treatment of effluents from sulfate pulp bleaching.

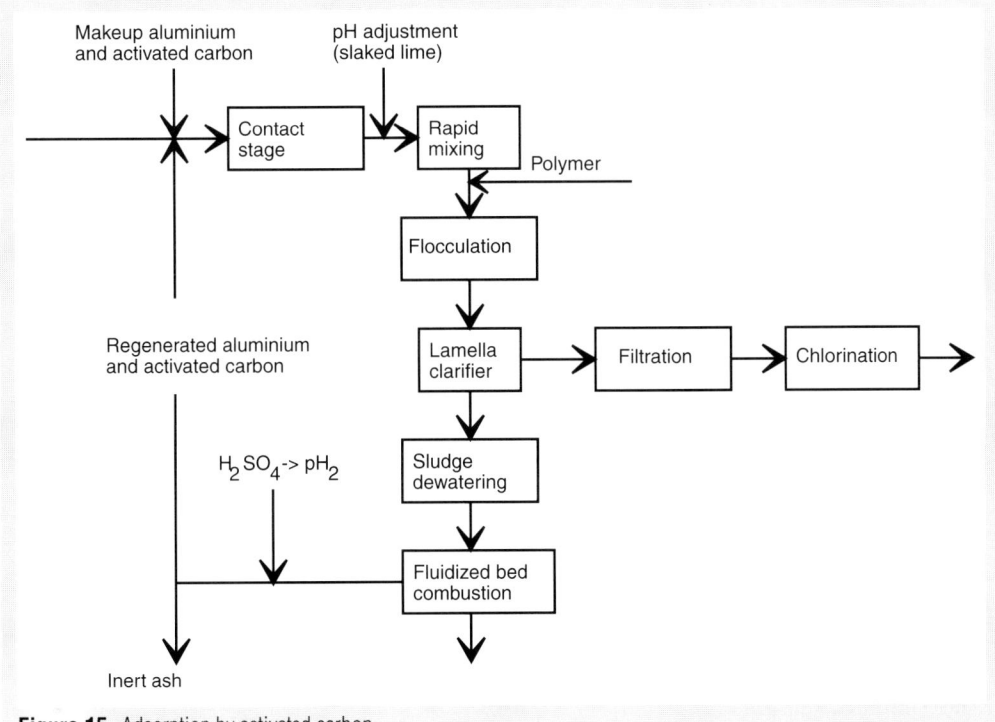

Figure 15. Adsorption by activated carbon.

3.2 Evaporation

The evaporation or distillation of effluents has usually been considered too expensive to be a realistic option.

However, evaporation is used for barking effluents and for some bleaching effluents in full scale. The method apperars to have good potential but more process and equiopment development will be needed.

CHAPTER 6

3.3 Lignin removal process

The lignin removal process (LRP) is based on the ability of acidified fiber sludge to cause organic material in effluent to precipitate on the surfaces of the fibers. The method has been used in full scale for effluents from non-wood pulp and paper mill.

The process consists of three main stages (Fig. 16):

- Acidification of fiber sludge
- Reaction stage
- Removal of fiber sludge from the effluent.

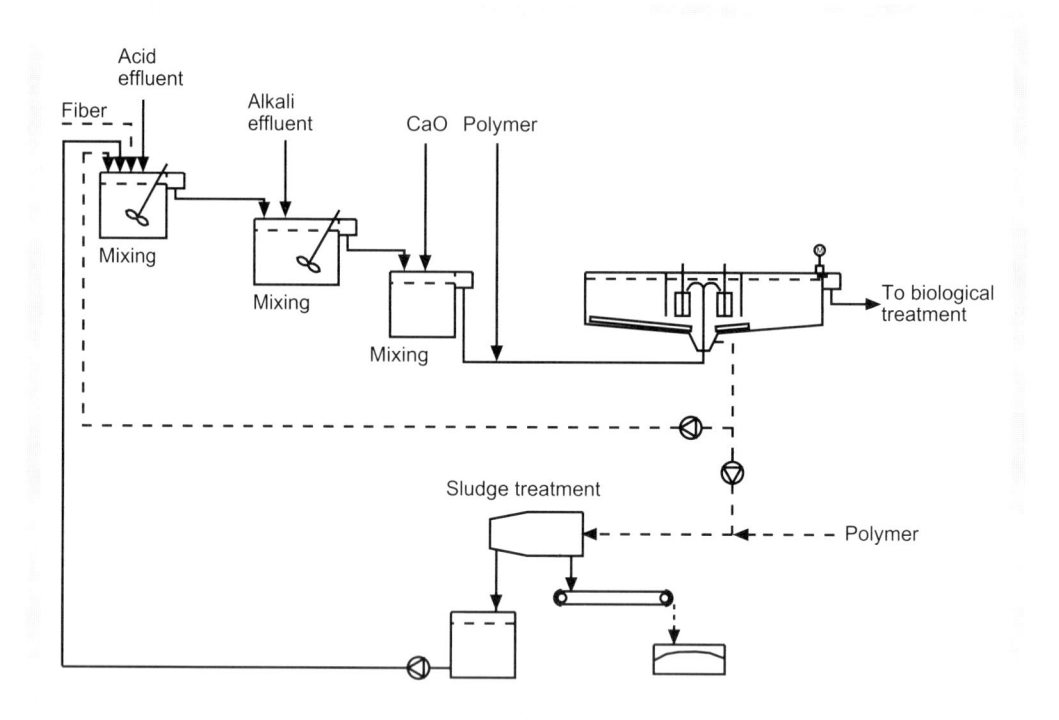

Figure 16. The LRP process.

The fiber sludge used in this process is acidified to bring the zeta potential of the sludge slurry close to zero. In the reaction stage, the acidified sludge is mixed with effluent. The result is precipitation of organic matter onto the fiber surfaces.

The method is cheaper than conventional chemical treatment. The sludge treatment is also more easy. The sludge can be reused for making of board or for concrete or tile products.

3.4 Stripping

Stripping is a way of removing volatile compounds from effluent. Heat and/or pH adjustment may be used to provide the right stripping conditions for the compounds in question.

Stripping takes place in a column. The effluent is introduced at the top and a gas (usually steam or air) at the bottom, so that the operation is counter-current. The effluent passes via a heat exchanger out of the base of the column.

The column may be packed with some inert material, or it may contain a variety of plate arrangements. Packed columns are used mainly for removal of malodorous sulfur compounds and have lower capacities. The packing particles are either plastic or stainless steel.

Columns with bell or aperture plate arrangements operate at high efficiency over a wide capacity range. Pressure loss is constant for a given steam loading.

In the forest industry, stripping is used for:

- Cleaning up contaminated condensates
- Removal of SO_2 from effluents
- Removal of NH_3 from effluents.

Stripping with air or steam is used to remove reduced sulfur compounds and volatile organics from contaminated condensates from chemical pulp digesters and evaporation units. At sulfite pulp mills, sulfur dioxide stripping is used to treat condensates before anaerobic treatment.

Stripping also can be used to remove ammonia from effluents in which the ammonia content is high.

3.5 Ion exchange

Ion exchange is a process whereby ions bound by electrostatic attraction to the functional groups of a solid-state compound are exchanged for ions in the surrounding solution. Ion exchange is used to remove specific ions from effluent.

An ion exchange process for the treatment of bleaching effluents was developed in Sweden in the early 1970s. The process removes colored compounds from the effluents very efficiently.

Effluent is led through a column packed with a suitable ion-exchange resin. When the resin becomes saturated with the ion concerned, it is regenerated with caustic soda followed by activation with sulfuric acid from the chlorine dioxide plant or effluent from the chlorination stage of the bleaching sequence.

The volume of eluent used to regenerate the column is about 5% of the volume of effluent treated. The eluate from the column can be combined with black liquor or led into the chemicals recovery system.

Ion exchange is used mainly for the treatment of bleaching effluents and for effluents with high COD loads.

3.6 Chemical oxidation

Organic compounds present in effluent can be oxidized either partially or completely by treatment with a suitable oxidizing agent. This results in a reduction of the total organic loading, while at the same time the organic material usually becomes more readily biodegradable.

The oxidizing agents used most often are:

- Chlorine
- Ozone
- Oxygen, air.

The process includes equipment for storing and dosing the oxidizing agent. In some cases, the oxidizing agent may be produced on-site. The effluent is allowed to react with the oxidizing agent for the required time. Residual unused oxidizing agent can be recovered from the aqueous or gaseous phase if necessary.

The use of chlorine obviously carries with it the problem that chlorinated organic compounds are formed.

3.7 Freezing

Pure water can be obtained from effluent by freezing. The idea is to obtain pure water for recycling back into the process. The following stages are involved:

- Heat removal from the effluent
- Ice crystal formation
- Removal of ice crystals from the concentrate
- Counter-current washing with clean water
- Raising the temperature of the ice
- Evaporation of concentrate.

The advantages of freezing over evaporation derive from the low temperatures used, which means the product is virtually free of volatile organics and there is less risk of corrosion. On the other hand, the efficiency of crystallization and subsequent removal of ice crystals depends on the suspended solids content of the effluent. This means that in some cases solids need to be removed before freezing.

In the forest industry, freezing is used to treat effluents from:

- Chemithermomechanical pulping
- Pulp bleaching
- Deinking of recovered paper

4 Removal of organic matter in effluent treatment

Effluent treatment is usually reported in terms of the percentage reductions achieved in certain parameters and the contents of various components remaining after treatment. Figs. 18 and 19 show the reductions in organic load achieved on treatment of effluents from mills producing conventional bleached pulp and wood-containing paper, respecti-vely, by means of clarifica-tion (without addition of chemicals) and activated sludge treatment. Clarification removes a high proportion of the waste, particularly in the paper mill case.

Nevertheless, the treated effluent carries substantial amounts of waste into the receiving water, although this is not reflected in its BOD load.

Typical results achieved with activated sludge treatment are presented in Table 1.

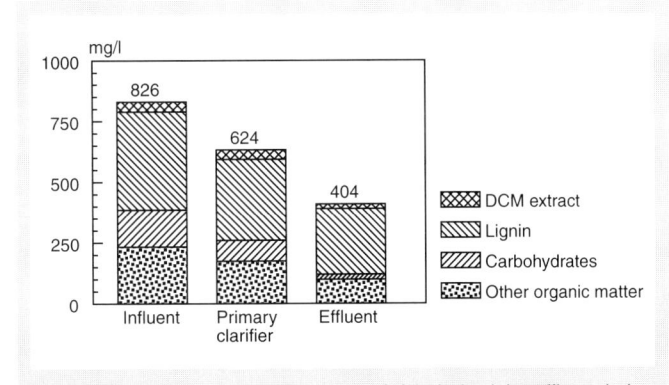

Figure 17. Reduction in organic content of chemical pulping effluent during treatment.

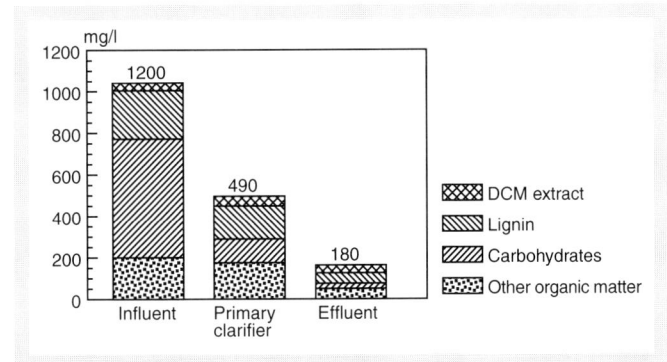

Figure 18. Reduction in organic content of effluent from wood-containing paper production during treatment.

Table 1. Effluent treatment by the activated sludge process.

	BOD removal, %	COD removal, %
Sulfate pulp mill	92–98	40–75*
Mechanical paper	92–98	70–90*
RCF-based paperboard	94–98	80–90
Effluent normally has a nitrogen content of 2–3 mg/l and a phoshorus content of 0.1–0.2 mg/l in solution.		

COD reduction depends on the bleaching method employed. Lowest reductions are for methods involving selective delignification.

CHAPTER 6

References

1. Virkola, N.E., Ed., *Production of Wood Pulp, Part II*, Turku, 1983, pp. 1399–1406.
2. Eckenfelder, W.W. Jr., *Industrial Water Pollution Control*, McGraw-Hill Book Co., Singapore, 1989, pp. 145–231.
3. Springer, A.M., *Industrial Environmental Control, Pulp and Paper Industry*, TAPPI PRESS, Atlanta, 1993, pp. 304–446.
4. Hynninen, P., and Ingman, L., Pulp Paper Intl. 40(11):63(1998).

Suggested reading

1. Schlegel, H.G., *General Microbiology*, Cambridge University Press, Cambridge, 1988.

2. Jörgensen, S.E., and Gromiec, M.J., *Mathematical Models In Biological Wastewater Treatment*, Elsevier Publishing, Delft, 1985.

3. Metcalf and Eddy, Inc., *Wastewater Engineering: Collection, Treatment, Disposal*, 2nd ed., McGraw-Hill, New York, 1979

CHAPTER 7

Reducing emissions to air

1	**Flue gases and the need for treatment**	**95**
2	**Ways of reducing particulate emissions**	**95**
3	**Ways of reducing sulfur dioxide emissions**	**99**
4	**Reducing nitrogen oxide emissions**	**101**
5	**Simultaneous removal of NO2 and SO2**	**102**
6	**Biological gas treatment**	**102**
7	**Collection and disposal of concentrated and dilute malodorous gases**	**102**
7.1	Concentrated malodorous gases	104
7.2	Dilute malodorous gases	104
	References	107

CHAPTER 7

Reducing emissions to air

1 Flue gases and the need for treatment

The different combustion processes used in the forest industry give rise to a variety of flue gases requiring different treatments. This chapter discusses current methods for cleaning up flue gases from industrial processes, with special focus on chemical pulping. The discussion includes citation of typical applications in conjunction with each method.

Information describing the combustion processes and the need to treat the different flue gases are in keeping with present-day requirements for air pollution control.

Process	Treatment needed
Energy generation	Particulates, SO_2, NO_x
Recovery boiler	Particulates, SO_2, TRS
Lime kiln	Particulates, TRS
Burning malodorous gases	SO_2
Bleaching and chlorine dioxide production	Cl_2, ClO_2, VOC (methanol, chloroform)
Others	VOC

The need to remove sulfur dioxide from flue gases arising from energy generation in the forest industry depends on the fuels used. Nitrogen oxides are controlled by combustion conditions..

Formation of reduced sulfur compounds (TRS) from recovery boilers are controlled internally via the process conditions. The need to remove sulfur dioxide diminishes considerably when black liquor with high dry solids is burned.

Process improvements – basically better washing of lime mud – reduce TRS emissions via lime kiln flue gases.

2 Ways of reducing particulate emissions

Numerous methods have been developed for removing solid particles from flue gases. Their efficiencies range from just over 50% removal up to more than 99.8% (Table 1).

CHAPTER 7

Table 1. Features of flue gas treatment equipment

Method	Material to be removed	Optimum particle size Ø μm	Efficiency (removal as % of incoming amount)	Space requirement[*]	Comments
Mechanical removal					
Separation chamber		>50	<50	L	Pre-treatment
Cyclone	Dust	5–25	50–90	M	
Momentum separation		>10	<80	S	
Filter		<1	>99	L	Impaired by moisture
Wet removal					
Spray tower	Dust and gases	25	<80	L	
Wet cyclone		>5	<80	L	
Impact scrubber		>5	<80	L	
Venturi		<1	<99	S	
Electrostatic precipitator		<1	95–99.5	L	
Gas scrubber	Dust and gases	>1% content	>90	ML	

[*] S = small, M = medium, L = large

The most important methods currently used in the wood processing industry to remove dust from flue gases are as follows, starting with the most efficient and widely used:

- Electrostatic precipitator
- Venturi scrubber
- Multicyclone.

All the above give dust removal rates of over 90%. This is essential for compliance with today's limits imposed on particulate emissions. Because of their high efficiency, electrostatic precipitators are becoming increasingly widespread.

The removal of dust particles in an **electrostatic precipitator** is based on the ionization of gas molecules. The flue gases are led through a strong electric field, which produces both negatively and positively charged ions. Most of the ions, however, are negatively charged, and these move toward the positive collecting plates. A small number of positively charged ions are formed, and these are attracted toward the discharge electrodes.

The ions formed in the electric field adhere to dust particles present in the gas, causing the dust particles to become similarly charged and to move toward either the collecting plates or discharge electrodes.

A mechanical rapping system dislodges the accumulated particles, which fall into collection hoppers or onto conveyors. The dust is removed from the hoppers either mechanically or pneumatically.

Reducing emissions to air

An electrostatic precipitator consists of the following:
- Ionization chamber
- Collection systems
- Discharge systems
- Gas distributors
- High-tension supply and other electrical equipment.

The ionization chamber frame is made from steel beams or reinforced concrete. The chamber has a gas inlet and exit and is equipped with ash hoppers. Some chambers can have flat bottoms, in which case a scraper is used to remove the collected ash. The chamber is thermally insulated.

Collection systems consist of collecting plates mounted on brackets, together with a rapping system for dislodging accumulated dust. The collecting plates are specially profiled for greater efficiency. The rapping system comprises a geared motor and a shaft driving a set of hammers. The frequency with which the hammers strike the plates can be controlled.

Discharge systems consist of discharge electrodes, supporting structures, and rapping systems. The electrodes themselves are made of steel strips attached to tubular frames. The frames, in turn, are mounted on supporting structures insulated from the ionization chamber and collection systems. The rapping arrangement is similar to that used in collection systems.

The purpose of the gas distributors is to ensure an even distribution of incoming gas across the entire surface of the precipitator. Profiled or perforated grills are often used.

The high-tension system comprises a transformer/rectifier, control cabinet, and possibly also remote controls.

In recent years, electrostatic precipitators have been introduced for dust removal from an increasing variety of flue gases. The technique has proved very effective for cleaning up flue gases from:
- Recovery boilers
- Lime kilns
- Bark-fired boilers
- Coal/peat-fired boilers.

Electrostatic precipitators are highly efficient, with dust removal rates generally around 99.5%–99.8%. The dust content of the air leaving the precipitator is normally less than 100 mg/Nm^3.

The venturi scrubber operates on the wet principle. Dust particles in the gas collide at high velocity with water droplets, and the resulting dust/droplets are subsequently removed from the gas in a cyclone.

CHAPTER 7

A two-stage venturi scrubber, in which the venturi itself is followed by a spray tower, is used fairly widely in the forest industry.

Venturi scrubbers can be positioned either vertically or horizontally. Water is sprayed into the gas stream before the throat. As the gas/water droplet mixture accelerates through the throat, the water forms very small droplets on which the dust particles form agglomerates.

Venturi scrubbers are most commonly used to remove dust from lime kiln flue gases. At the same time, sulfur dioxide in the flue gases is absorbed into the scrubbing solution. Venturi scrubbers have also been used to remove dust from recovery boiler flue gases prior to sulfur dioxide removal.

Venturi scrubbers generally operate at efficiencies of 95%–99%, depending on the arrangement (single or two-stage scrubbing), which is almost as good as an electrostatic precipitator. Recently, however, venturi scrubbers have started to be replaced by electrostatic precipitators.

A multicyclone is a dynamic separator in which dust removal is achieved with the help of centrifugal force. The dust-laden gas usually enters the cyclone units axially. The dust particles are flung against the walls and are drawn down into the dust hopper. Multicy-

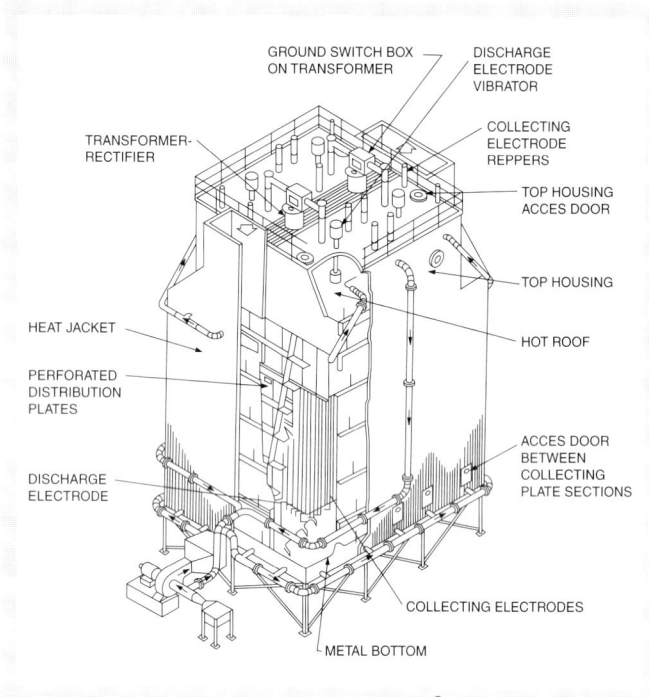

Figure 1. Structure of an electrostatic precipitator[2].

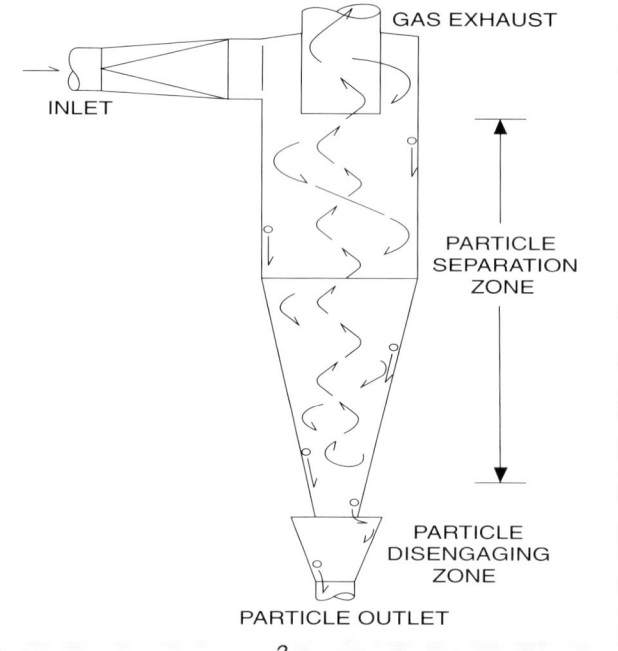

Figure 2. Principle of the cyclone[2].

clones have the advantage that, in certain gas velocity ranges, maximum use is made of the centrifugal force, thus giving fairly high dust removal rates.

A multicyclone consists of a number of small-diameter cyclones into which the gas is fed equally.

Multicyclones are used to treat flue gases from bark-fired or multi-fuel boilers. Dust removal rates are usually around 90%, with exhaust gas dust contents reduced down to 300 mg/Nm^3. In many cases, however, multicyclones are unable to meet today's limits for particulate emissions and are being replaced by electrostatic precipitators.

Numerous other dust removal systems have been used, or are still being used, in the forest industry. Some of these are shown in Fig. 3.

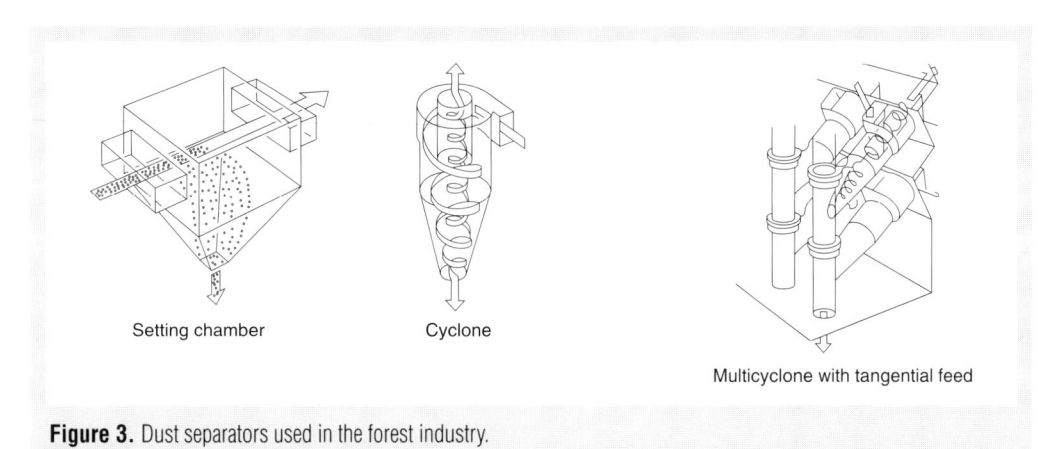

Figure 3. Dust separators used in the forest industry.

3 Ways of reducing sulfur dioxide emissions

The key to managing sulfur dioxide emissions is process control, which maintains a suitable chemicals balance. Modifying the process conditions – for example, evaporating black liquor to a higher dry solids content – can reduce emissions. Using gas scrubbers also can reduce sulfur dioxide emissions.

Fig. 4 shows an old simplified scheme illustrating the sulfur balance for a sulfate pulping process (OABCD added chemicals, DEEO chemical losses). The reduction in sulfur emissions ach-ieved, mainly by cutting sulfur losses during washing and chemicals recovery, means that less sulfur now has to be introduced into the process. Collection and burning of malodorous gases, including sulfur

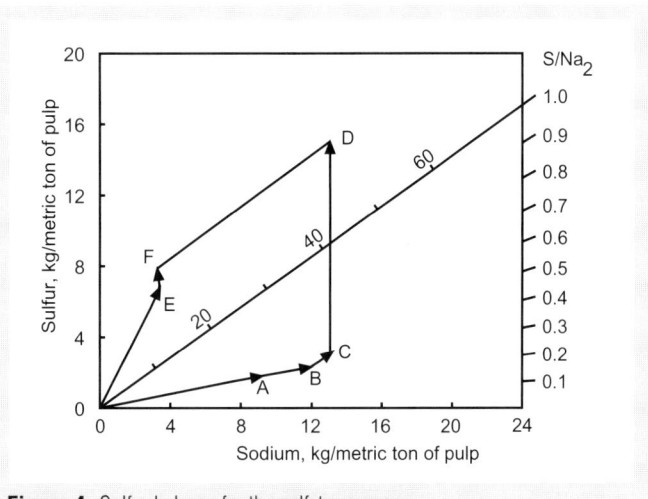

Figure 4. Sulfur balance for the sulfate process.

CHAPTER 7

dioxide recovery, is having the same kind of effect. Reducing the need for sulfur has gradually ousted Glauber's salt from its traditional role as makeup chemical in the sulfate process. With modern chemicals recovery systems, managing the chemicals balance cannot be overlooked as a means of cutting sulfur emissions.

Today, various types of scrubber (Fig. 5) are used to remove SO_2 from flue gases in the forest industry.

In their own energy generation, it has to be assumed that forest industry plants (paper mills and sawmills) can exert some control over sulfur dioxide emissions by burning low-sulfur fuels. Such fuels include wood waste, bark, peat, and natural gas, as well as a certain amount of low-sulfur oil. In the generation of energy on a national scale (coal-fired power plants), the following techniques cut sulfur emissions:

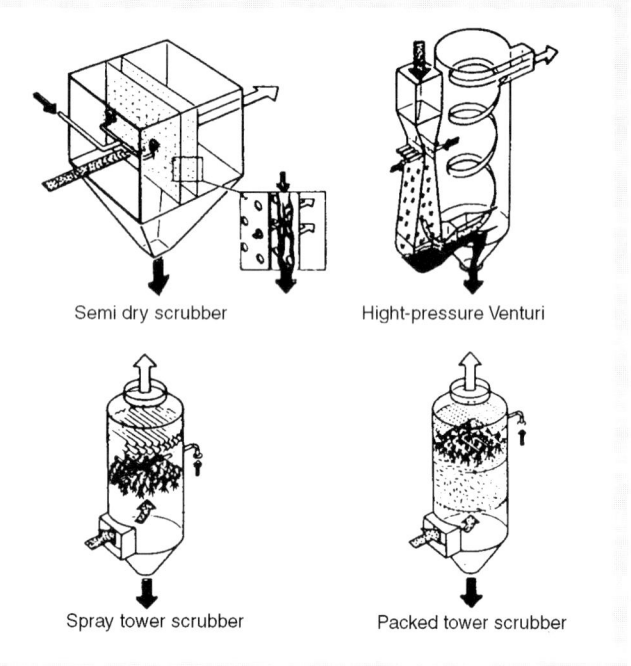

Figure 5. Scrubbers used in the forest industry.

- Dry injection (CaO injected dry into the boiler): sulfur removal 70%–80% (Fig. 6)
- Semi-dry method: sulfur removal 90%–95%
- Wet method: sulfur removal 90%–95%.

Sulfur dioxide formed during the combustion processes employed at pulp mills is recovered by scrubbing and returned to the chemicals recovery cycle.
Briefly, the <u>flue gas scrubber</u> works like this:

- The scrubber is a closed unit into which the gas containing impurities runs in a counter-current motion to the scrubbing solution so as to achieve maximum contact between the two.

At sulfate pulp mills, the scrubbing solution usually is either NaOH or oxygenated white liquor.

Several different types of scrubber are in use (see also Fig. 5):

The reader can find an earlier discussion of the <u>venturi scrubber</u> under "Ways of reducing particulate emissions." Sulfur dioxide removal takes place largely in the spray tower after the actual venturi. The advantage of the venturi scrubber is that it removes both particulates (venturi section) and sulfur dioxide (in two stages) all in the same unit.

The sulfur dioxide removal achieved is usually in the region of 80%–90%, depending on the design of the unit.

The fact that sulfur dioxide removal efficiency is not especially high explains why venturi scrubbers are not widely employed for this purpose.

<u>Packed column scrubbers</u> are highly efficient in removing sulfur dioxide as the area of contact between gas and scrubbing solution is much greater than in the venturi arrangement. In a packed column scrubber, spraying scrubbing solution in at the top through several jets uniformly covers the packing granules. Feed the gas in at the base of the column, possibly after removal of particulates in a venturi or spray tower. In the latter case, there is a need for two separate liquid cycles to prevent the packed column from getting clogged up. Adjust the pH of the scrubbing solution through the introduction of fresh alkali and its concentration through the addition of water. The scrubbing solution cycle also includes a heat exchanger to allow the temperature of the solution to be adjusted to optimum for sulfur dioxide adsorption.

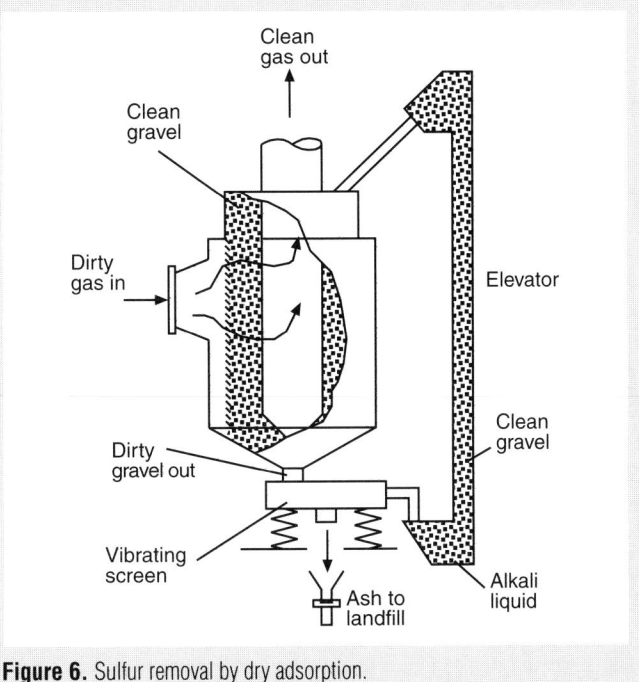

Figure 6. Sulfur removal by dry adsorption.

The SO_2 removal efficiency of a packed column scrubber is usually about 95%, but higher if the system consists of several stages. Such scrubbers are frequently used to remove SO_2 from the flue gases from recovery boilers; more recently they have been introduced for SO_2 recovery from the flue gases from the incineration of malodorous gases.

4 Reducing nitrogen oxide emissions

Due to the process conditions, nitrogen emissions from pulp mills are low, especially those from recovery boilers. Nitrogen emissions are naturally higher if the fuels used have higher nitrogen contents. Nitrogen emissions are high from mills employing the ammonium bisulfite process.

Nitrogen oxides can be removed from flue gases, principally those from auxiliary boilers, using either catalytic or noncatalytic reduction. Both processes involve the use of a solution or gas containing either ammonia or urea.

Nitrogen oxides can be reduced to molecular nitrogen and water. The amount of ammonia used must be sufficient but not excessive, since any unreacted ammonia passes through with the flue gases into the air. The ammonia requirement is very close to the stoichiometric requirement.

Selective catalytic reduction (SCR) is used mainly for treating power plant flue gases. The SCR method has not been adapted commercially for recovery boiler flue gases, as the sodium sulfate dust emitted with the recovery boiler flue gases is known to poison the commercial catalysts employed.

Selective noncatalytic reduction (SNCR), on the other hand, is suitable for the removal of nitrogen oxides from recovery boiler flue gases. Reductions of around 60% are normally achieved[1].

Other methods of nitrogen oxide reduction from recovery boiler flue gases include the use of tertiary air. This has been shown to reduce emissions by some 30%. Treating the gases with reducing scrubbing solutions can provide removal rates of up to 50%–60%[1].

5 Simultaneous removal of NO_2 and SO_2

Methods for the simultaneous removal of sulfur dioxide and nitrogen oxides are currently under development. Most advanced in this respect are methods based on active carbon and electron irradiation.

Although these methods are still under development, some pilot plant tests have been made and some power plant applications tried out. Both methods carry potential for nitrogen removal in the future.

6 Biological gas treatment

Biological treatment involves the use of micro-organisms to fix and break down gaseous impurities.

Biological treatment is suitable for cleaning up gases from effluent treatment plants, composting, and the pharmaceutical, food, and printing industries. However, the method is at present inadequate for dealing with malodorous gases from pulp mills.

7 Collection and disposal of concentrated and dilute malodorous gases

Sulfate pulping gives rise to malodorous gases containing hydrogen sulfide and reduced organic sulfur compounds. These volatile compounds can be present at high concentrations in certain off-gases and condensates from digesters and evaporators. These are referred to as concentrated malodorous gases. Dilute malodorous gases, on the other hand, are collected from a variety of points on the fiber and cooking liquor lines. While the concentrations are low, these gases often constitute large volumes. The concentration of offending compounds in the concentrated gases usually exceeds 10%, while that in dilute gases is measured in parts per thousand.

Reducing emissions to air

Table 2. Malodorous gases from sulfate pulping.

Source of smell	Amounts	
	m³/metric ton of pulp	kg S/metric ton of pulp
Concentrated gases	5–15	0.4–0.8
Blow gases, batch digesters	1–3	0.1–0.2
Relief gases, batch digesters	0.5–1.5	0.1–0.4
Blow gases, continuous digesters	0.5–1.5	0.1–0.4
Uncondensed gases from evaporators	1–10	0.4–0.8
Steam stripping of contaminated condensates	15–25	0.5–1.0
Flue gases		
From lime kiln	800–1 500	0.2–0.8
From recovery boiler		
(dry)	6 000–8 000	
(wet)	9 000–12 000	
- modern recovery boiler	0.05–0.10	
- after direct contact evaporation	3–4	
Dilute gases		
Exhausted from pulp washing		
- suction filter washing	1 000–1 500	0.05–0.1
- pressure filter washing	1–2	0.01–0.05
- diffuser	1–2	0.01
Exhausted from tall oil production	2 000–3 000	0.1–0.2
Storage tank vapors (black liquor, contaminated condensates)	20–30	0.5–0.2
Solvent vapors	500–1 000	0–0.05

The smell associated with malodorous gases is due to their content of reduced sulfur compounds, which have a very low odor threshold.

Table 3. Amounts of malodorous compounds and biodegradable substances formed and released during sulfate pulping.

Compound	Amount
SULFUR-CONTAINING COMPOUNDS	kg S/metric ton of pulp
Hydrogen sulfide	0.5–1.0
Methylmercaptan	
Dimethylsulfide	1.0–2.0
Dimethyldisulfide	
NON-SULFUR COMPOUNDS	kg/metric ton of pulp
Methanol	6–13
Ethanol	1–2
Turpentine[*]	4–15
Guaiacol	1–2
Acetone	0.1–0.2

[*] During pine cooking, most is recovered.

CHAPTER 7

Effective odor suppression requires that all malodorous gases are collected in a closed system and burned.

Concentrated and dilute malodorous gases are collected in separate systems.

Gases are usually transferred from one place to another using vapor ejectors. The pipes used are fitted with equipment and devices necessary to prevent explosions, fractures, and fire, as well as with traps for condensation. To ensure continuity of operation, two separate combustion points are required. The sulfur dioxide formed when the gases are burned is collected and returned to the chemicals recovery system.

7.1 Concentrated malodorous gases

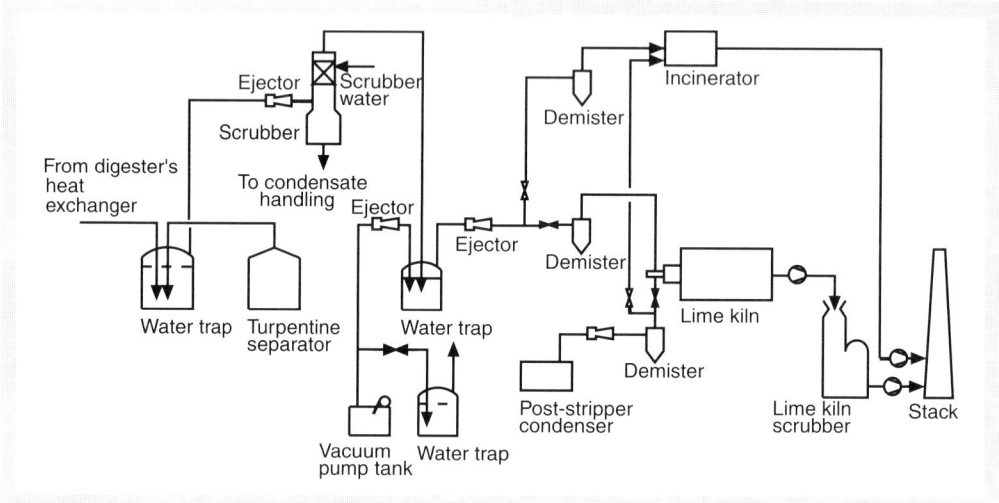

Figure 7. Collection system for concentrated malodorous gases.

In the sulfate pulping process, concentrated malodorous gases are released from the digesters and the evaporators. Although the volumes are relatively small, the sulfur content is high (see Table 3). Fig. 7 illustrates the way in which these gases are collected.

In most cases, gases from stripping are combined with concentrated malodorous gases for burning.

7.2 Dilute malodorous gases

Dilute malodorous gases originate during the handling of black liquor and the production of tall oil. To overcome the potential odor problem, all such gases must be collected, which means also catering for sources such as sewers. A system for collecting dilute malodorous gases is shown in Fig. 8.

The gases can be burned in a separate combustion unit or else combined with some other fuel and burned in, say, the recovery boiler or lime kiln. Combustion requires a sufficiently high temperature and adequate burning time exact values depending on the burner type.

Reducing emissions to air

Figure 8. Collection system for dilute malodorous gases.

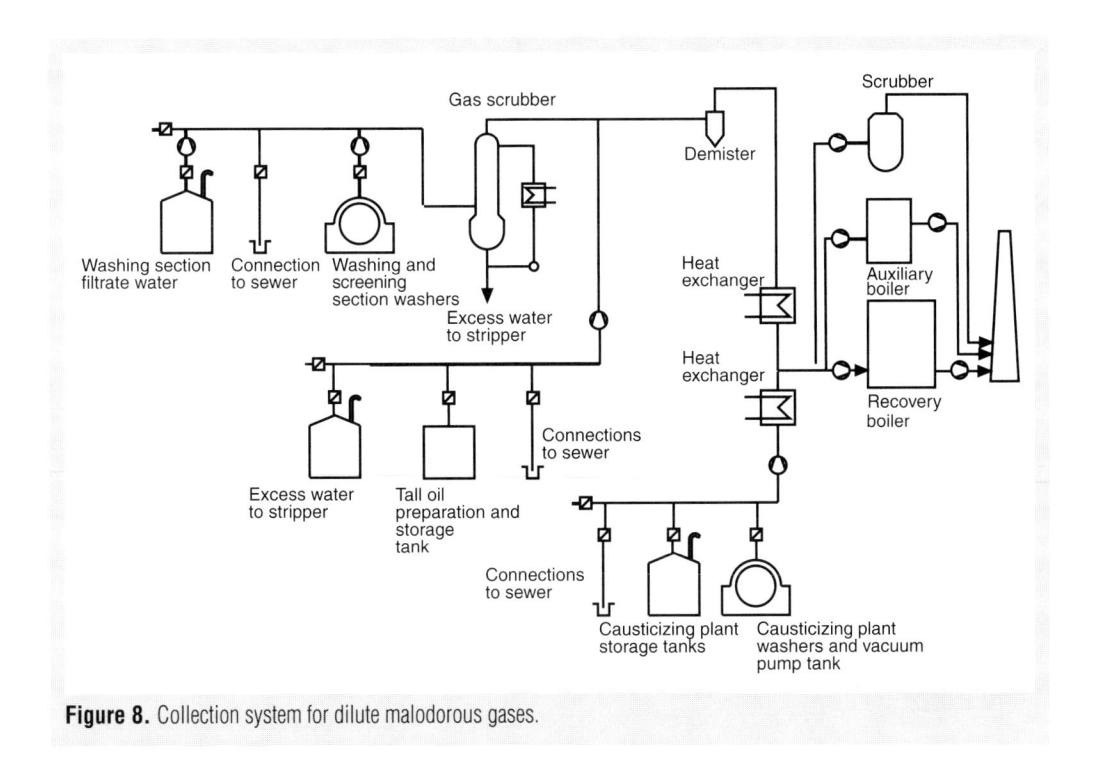

The combustion unit can be followed by a heat exchanger or steam boiler so that some of the energy used to burn the gases is recovered as hot water or steam. This naturally makes the entire disposal system more economical.

The malodorous sulfur compounds produced during sulfate pulping can be burned and some energy recovered as described above. Concentrated gases may first be mixed with some dilute gases. Dilute gases are generally burned in the recovery boiler, into which they are fed along with the tertiary air.

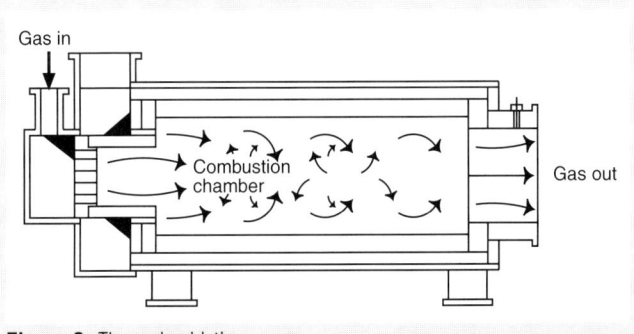

Figure 9. Thermal oxidation.

In thermal oxidation (i.e., combustion), reduced sulfur compounds are oxidized to sulfur dioxide. This does not reduce the emission, but merely changes its chemical structure. After thermal oxidation, sulfur dioxide is often removed from the gases before they are released into the atmosphere.

Some 90% of Finland's sulfate pulp mills employ thermal oxidation. The gases burned are mainly concentrated malodorous sulfur compounds. At about one-third of

CHAPTER 7

sulfate mills, dilute gases are dealt with in the same way; a few mills mix these with tertiary air going to the furnace of the recovery boiler.

Table 4. Availability (%) of odour control equipment.

	"Normal"	Target
Collection and disposal of concentrated malodorous gases	95–98	100
Collection and disposal of dilute malodorous gases	60–80	95–100
Condensate stripping*	90–95	95–100

*Availability increased by liquefying methanol.

As in thermal oxidation, the catalytic oxidation of gaseous sulfur compounds results in the formation of sulfur dioxide, carbon dioxide, and water vapor.

Catalytic oxidation is suitable for dealing with malodorous gases from sulfate pulping.

Although catalytic oxidation has been studied for this purpose, there are as yet no commercial applications. At present, catalytic oxidation is used to deal with solvents and other volatile organic compounds.

The biggest problems currently encountered in odor prevention at sulfate pulp mills include the following:

- No reserve place is available for burning concentrated gases, or else there are delays in getting reserve combustion units operational.

- There are no action plans to cater for escape of malodorous gases due to problems with condensate stripping.

- The process hardware does not lend itself to the collection of dilute malodorous gases.

- Apart from the recovery boiler, there is no other way of burning all dilute malodorous gases. Burning capacity of the other boilers and lime kiln are too small. See also Table 4.

References

1. Kwaerner Pulping, Information brochures, Tampere, 1997.
2. Springer, A.M., Industrial Environmental Control, Pulp and Paper Industry, TAPPI PRESS, Atlanta, 1993, pp. 583–591.

CHAPTER 8

Solid and liquid wastes

1	**Waste volumes from different production processes**	**109**
1.1	Sludges	109
1.2	Ash	111
1.3	Other wastes	111
1.4	Amounts of waste per product	111
2	**Waste handling**	**113**
2.1	Sludge handling	113
3	**Waste incineration**	**123**
3.1	Combustion of sludge	123
4	**Landfilling of wastes**	**125**
5	**Landfilling of sludges**	**125**
6	**Disposal of other wastes**	**127**
6.1	Soil improvement	127
6.2	Returning sludge to the production process	129
6.3	Production of animal feeds from sludge	129
6.4	Other products from sludge	130
	References	131

CHAPTER 8

Solid and liquid wastes

Introduction

Different processes within the forest industry result in the formation of different solid and sludge-like wastes. In terms of volume, the biggest are those from the treatment of effluents and flue gases, although wood waste is also produced in large quantities. Other wastes arising at pulp mills are green liquor, lime sludge, and other process wastes. Wastes arising from paper production include those containing fillers and additives. Chemimechanical pulping and deinked pulp production give rise to their own characteristic wastes.

Increasing restrictions on emissions have resulted in an increase in the volume of waste produced by forest industry plants. This, together with the difficult physical form of these wastes, poses problems in waste handling and disposal. The biosludge formed during biological effluent treatment is particularly problematic. Sludges with low dry solids contents have to be conditioned before they can be properly handled. Such sludges are usually thickened, raised to high dry solids contents using a belt filter press, and then either burned in a bark-fired boiler together with bark from wood handling, or used for landfill. Biosludge is invariably burned to render it easier to handle, as seldom is energy production from biosludge really worthwhile. Nevertheless, burning sludge is generally an economically viable option as it reduces the costs of landfill placement.

The problems associated with the landfilling of sludges and other wastes are the large volume involved and the possibility of hazardous substances getting into the environment. Also, landfill sites are not always up to the standard required, and sites are becoming more expensive and harder to find. This means finding alternative means of disposal or at least treating wastes before landfilling to minimize their possible adverse effects. The most promising uses for sludges and ash include soil improvement, road building, brick industry, forestry and horticulture.

1 Waste volumes from different production processes

In the forest industry, the greatest accumulations of waste consist of sludges, ash, and wood waste.

1.1 Sludges

The treatment of forest industry effluents produces primary sludge from mechanical treatment and biosludge from biological treatment. In addition to these, chemical sludges are produced from the treatment of coating effluents and also when mechanical treatment is boosted with chemicals. According to data gathered by the Finnish Forest Industries Federation, a total of 680 000 metric tons (wet weight) of sludge was

CHAPTER 8

produced in Finland in 1992. Most of this was primary sludge.

The volumes of primary sludges produced vary greatly according to the product being manufactured. Primary sludge represents about 2% of the total output of wood-pulp and paper as follows :

- Mechanical pulping, 15–20 kg/metric ton (including bark sludge)
- Chemical pulping
- Sulfate pulp, 20–25 kg/metric ton
- Semichemical pulp, 25–30 kg/metric ton
- Paper and board production, 5–10 kg/metric ton.

At modern pulp mills, sludge formation corresponds to about 1% of pulp production.

Biosludge (including suspended solids with efluent from the primary clarifier) is formed at the rate of 10–20 kg/metric ton of pulp or around 0.2–1.2 kg/kg of BOD removed, depending on the treatment method used. At paper mills, the BOD load to the treatment plant is usually much smaller than at pulp mills, and the amounts of biosludge formed are therefore smaller. The Finnish forest industry produces biosludge at the rate of about 50 000 metric tons/year (dry weight).

Chemical treatment plant sludge is produced at an estimated rate of 8 000 metric tons/year (dry weight).

Deinking plants give rise to roughly 150 kg of sludge per metric ton of deinked fiber. In 1991, the total produced was an estimated 33 800 metric tons.

Table 1. Sludge formation calculated in metric tons dry weight.

	1991	1995	2000	2005	2010
Scenario A					
Primary sludge, tot.	255 000	259 000	267 000	279 000	298 000
Fiber sludge	201 000	201 000	206 000	215 000	230 000
Bark sludge	54 000	58 000	60 000	64 000	68 000
Biosludge, total	465 500	56 700	63 300	64 900	68 000
From pulp mills	23 200	26 600	26 900	27 600	28 200
From paper mills	23 300	30 100	36 400	37 300	39 700
Deinking sludge	33 800	44 500	41 200	38 000	38 700
TOTAL	335 000	360 000	371 000	382 000	405 000

1.2 Ash

The biggest volumes of ash produced result from energy generation and flue gas treatment. The amount and characteristics of the ash depend on the fuel and the combustion technology used. The total amount of ash formed in Finland in 1992 was 255 000 metric tons, most of which was used for landfill. Some 60 000 metric tons was used as forest fertilizer.

Ash formation has been estimated[1] per metric ton of product as follows: for sulfate pulp production 6 kg/a.d. metric ton, mechanical pulp production 3 kg/a.d. metric ton, and deinked pulp over 6 kg/a.d. metric ton. Ash formation attributable to papermaking varies greatly depending on the pulp used, ranging from 3 kg to over 10 kg/metric ton (including ash due to production of the necessary pulp).

1.3 Other wastes

<u>Wastes from wood processing</u>
In 1992, the Finnish pulp and paper industry gave rise to 2 080 000 metric tons of wood waste (mainly bark), of which over 95% was used for energy generation and pulp production. The remainder was trash, stones, and other miscellaneous wastes, which were used for landfill.

<u>Wastes from chemicals recovery</u>
The preparation of cooking liquor for pulping results in the formation of up to 10 kg of green liquor dregs per metric ton of pulp. These dregs are usually taken to the landfill site. Preparation of white liquor also produces some lime sludge. Although this has been used to neutralize effluents prior to treatment, it is normally taken for landfill.

<u>Waste from papermaking</u>
Papermaking gives rise to wastes containing additives and fillers such as pigments, as well as paper waste.

Other wastes generated by the forest industry include:

- Problem wastes (solvents, oils, preservatives, and used capacitors, accumulators, and batteries)
- Mixed industrial wastes such as scrap metal, and building and packaging waste
- Waste from offices and laboratories.

1.4 Amounts of waste per product

The simulation model[1] has been used to estimate the amounts of waste produced by the forest industry in 1995 for each type of product.

The waste produced in the case of bleached softwood sulfate pulp was about 45 kg/a.d. metric ton and in the case of hardwood pulp around 50 kg/a.d. metric ton. For

CHAPTER 8

both of these products, the amount of organic waste was easily the highest at around 30 kg/a.d. metric ton.

Waste from stone groundwood production has been calculated to be about 13 kg/a.d. metric ton, for pressure groundwood about 15 kg/a.d. metric ton, for TMP about 18 kg/a.d. metric ton, and for semichemical pulp about 30 kg/a.d. metric ton. For these pulps, biosludge in proportion to total waste was by far the highest.

Fig. 1 shows the amounts of solid wastes produced per metric ton of product for different pulping processes. As can be seen, deinking results in large amounts of solid waste (around 150 kg/a.d. metric ton).

Pulp and paper mills are producing other solid and liquid wastes like metals, waste chemicals, waste paper and packaking wastes. These are treated according local waste management plans but not normally included to the waste loading kg/a.d metric ton.

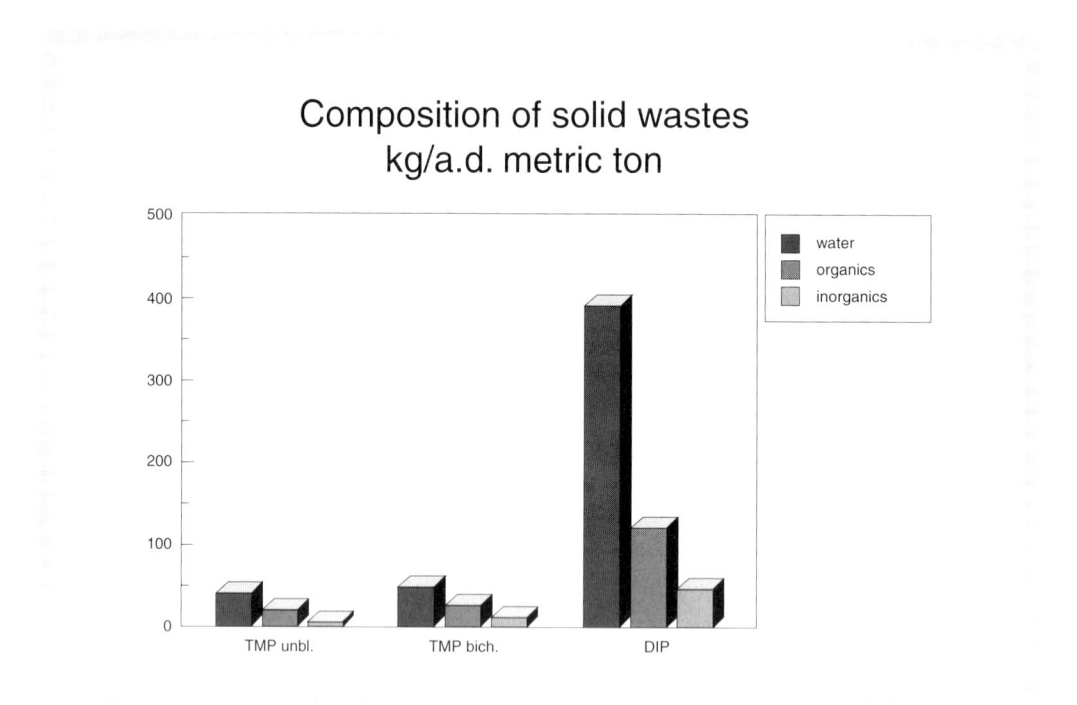

Figure 1. Formation of solid wastes from production of chemical pulp, mechanical pulp, and deinked recycled fiber pulp.

Calculated per paper grade (including waste attributable to pulp production), the amount of waste due to papermaking varies from about 25 kg/metric ton for SC paper to almost 60 kg/metric ton for bleached chemical pulp board. The amount of inorganic waste is highest for bleached chemical pulp board and kraftliner, and lowest for newsprint. In the case of tissue, some 20 kg/a.d. metric ton of the total waste of around 47 kg/a.d. metric ton is sludge from effluent treatment.

Table 2 shows predicted waste quantities (Scenario A) calculated by simulation[1]. The small sludge quantities shown in the table should be treated with some caution. The sludge quantities presented are smaller than those found in practice (see Table 1).

Table 2. Trend in waste volumes (metric tons/year).

	1991	1995	2000	2005	2010
Solid wastes					
Ash	127 000	150 000	162 000	173 000	183 000
Fiber reject	19 000	21 000	22 000	23 000	25 000
Dregs	106 000	119 000	122 000	139 000	146 000
Sludge from effluent treatment, dry	136 000	130 000	136 000	141 000	150 000
Deinking sludge, dry	33 800	36 500	46 300	51 100	56 400
Sludge water	303 000	318 000	345 000	363 000	389 000
TOTAL	725 000	773 000	833 000	890 000	950 000

2 Waste handling

General

As already mentioned, forest industry wastes need to be reused because of the shortage of new landfill sites and because the wastes can cause problems when landfilled.

The most difficult wastes to handle are sludges from effluent treatment. It is not that hazardous substances cause problems in final disposal but rather that the physical form of the wastes is difficult and that the volumes involved are large. Landfill disposal of sludges will become more expensive in the future, and may even be banned completely in some areas. Because of their low dry solids contents, sludges need to be processed before being reused or disposed of by landfilling or burning. The most widely used methods for processing sludges are presented in the next section.

Handling other wastes produced by the forest industry is not a major problem. Ash is generally disposed of by landfill. Normally, ash only has to be wetted before disposal to prevent dust release. The dregs from chemicals recovery at pulp mills are usually pressed and also used for landfill.

2.1 Sludge handling

The liquid effluents from forest industry processes contain many different components. Fiber sludge contains fibers and fiber fragments, bark, fillers, and grit. Biosludge, on the other hand, is composed of both dead and living bacterial cells. Chemical sludges contain flocs formed by the coagulants used.

The aim in sludge dewatering is usually to achieve as high a dry solids content as possible, as this assists subsequent handling. In a sludge suspension, water exists in the following forms:

- Free water
- Capillary water
- Bound and intercellular water.

CHAPTER 8

Free water can be removed simply by gravity settling. Much of the water is removed by allowing the sludge to stand, when solids sink to the bottom and water forms a supernatant. Capillary water can be removed mechanically by filtration or centrifugation, processes that can be made more effective by first conditioning the sludge. Bound and intercellular water can be removed by drying, i.e., through evaporation.

Sludges can be characterized in terms of parameters such as:

- Dry solids content (evaporation)
- Solids content (filtration)
- Ignition residue
- Recovery rate and specific drainage resistance
- Fiber length distribution
- Ash content
- Pitch content
- Viscosity and pH
- Settling rate
- Zeta potential
- Freeness, compressibility
- Heat value, fiber analysis.

The nature of the sludges formed by the mechanical and chemical treatment of pulp and paper mill effluents depends on the type of production process involved, the raw material, water consumption, solids losses, and the nature of each solids fraction. Sludge can be characterized in terms of, say, its particle size distribution and its ash content, which reflects the filler content of the solids.

Sludge handling methods

The process of sludge dewatering can be divided into the following parts:

- Clarification
- Thickening
- Sludge conditioning/preliminary dewatering
- Mechanical water removal
- Further handling such as pressing and heat drying.

Fig. 2 shows the different stages of sludge handling and the different options available. A typical approach would be to thicken the sludges, either singly or together, dewater the mixed sludges in a belt filter press, and finally burn the sludge in a bark-fired boiler.

Solid and liquid wastes

Figure 2. Different stages of sludge handling.

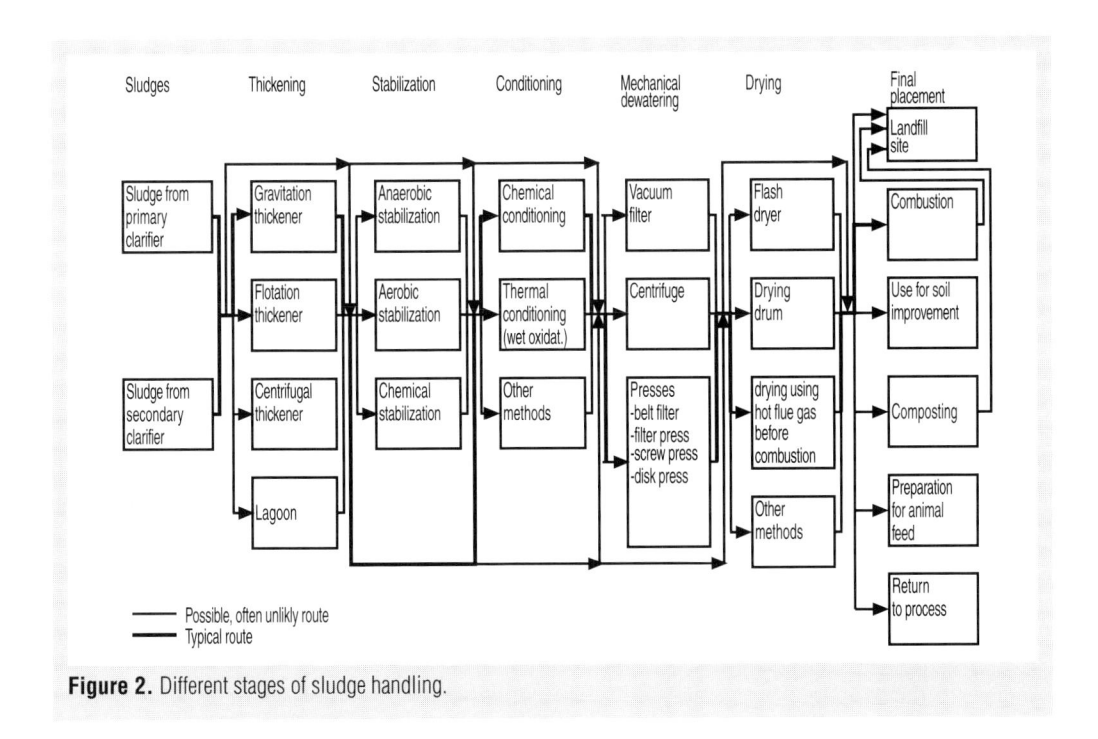

Sludge thickening is normally achieved by sedimentation. By this means, the solids content of the sludge is raised from 0.3%–3% up to 2%–10% (in the case of biosludge up to 1.5%–3%). Gravity thickening takes place in a circular tank resembling a clarifier, although the high solids load means the equipment has to be more robust and often includes scrapers. Tanks of smaller area are also sometimes used. Gravity thickeners are designed on the basis of sludge surface load theory. Thickeners are also provided with storage space for thickened sludge. Other types of thickener include dewatering drums, curved screens, and flotation thickeners. These are often designed on the basis of pilot tests.

Sludge conditioning may take place chemically or by means of heat. Today, only chemical conditioning is widely used. The best results are usually obtained with a dual system consisting of an inorganic salt and a polyelectrolyte (e.g., iron(III) sulfate and PAA).

The chemicals are usually fed into a flocculation tank with a 1–3 minute residence time to allow the chemicals to react. With certain types of dewatering equipment, some water has to be removed at this stage. This can be done using a rotating drum, from which the sludge discharged has a consistency of 6%–10%. The drum can also be fitted with a screw to raise the discharge sludge consistency to 10%–15%.

Biosludge arriving for thickening is still biodegrading fairly rapidly. The prevention or slowing down of this is referred to as stabilization. Stabilization is used mainly in conjunction with municipal sludges, but may also be employed in the forest industry.

CHAPTER 8

Stabilization methods include:

- Lime stabilization, whose effect depends on the amount of lime added. Stabilization with lime results in an increase in the amount of sludge. Dewatering properties improve.
- Anaerobic stabilization, a biological method carried out in a closed, oxygen-free environment.
- Aerobic stabilization, a biological method in which the sludge is aerated. The result is a lower dry solids content and better dewatering properties.
- Heat has been used at large municipal treatment plants to stabilize sludge and improve its dewatering properties.

Mechanical dewatering employs equipment based on filtration, centrifugation, pressing, or combinations of these. Filters are used to raise sludge consistency from 2%–5% up to 15%–30%. Although filters have good solids retention, they are susceptible to clogging if the sludge has a high pitch content.

Presses usually operate at either low pressure or high pressure. For high-pressure presses, the incoming sludge usually has to have a high solids content (8%–12%). Depending on the type of press employed, the final solids content is 25%–45%. Among low-pressure presses, the popularity of the belt filter press has increased as technology has advanced.

Mechanical dewatering equipment

Centrifuges

The centrifuge most commonly used in the wood-processing industry is the solid bowl decanter. This consists of a rotating bowl, inside which a screw conveyor rotates in the same direction. Sludge is fed in via the hollow screw shaft and from there into the bowl. The bowl rotates at roughly 1 500–4 000 rpm and the screw conveyor 2–50 rpm more slowly. The centrifuge controls consist of the speed of the bowl, the speed differential between bowl and conveyor screw, and the extent of solids recovery.

Sludge is fed into the center, where centrifugal forces press it against the bowl wall. The screw conveyor carries the compacted sludge along the wall toward the narrow end of the bowl, from where it is discharged. Water flows out as an overflow from the other end (Fig. 3).

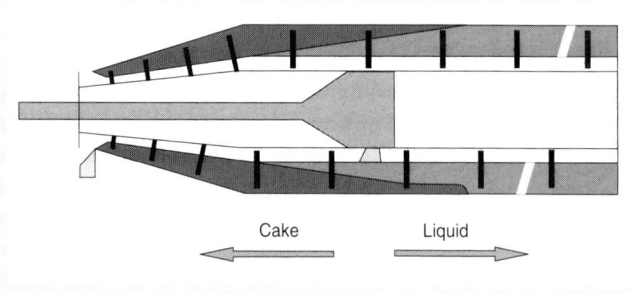

Figure 3. Decanter centrifuge[2].

Solid and liquid wastes

To improve dewatering efficiency, the sludge is first conditioned with polyelectrolytes.

The main use for the centrifuge today is in the dewatering of inorganic sludges, although it has been used for thickening biosludge. In practice, dry solids contents of 15%–30% can be obtained in this way. Centrifuges give a solids recovery of between 80 and 95%, depending on the type of sludge and the conditioning chemicals used. Centrifuges are not at their best with organic sludges; their energy consumption is also fairly high.

Vacuum filter

Vacuum filters (belt filters) are still used in big numbers by the forest industry worldwide, although in Finland they are no longer employed for dewatering effluent sludges. On the other hand, vacuum filters are useful for dealing with fiber-containing sludges. Fiber recovery and obtained dry solids are normally good. However, vacuum filters are not suitable for biosludge or chemical sludge.

Belt filter press

In a belt filter press, water is removed from the sludge in two stages: first by gravity and then by pressing (usually in several consecutive stages). The press is controlled by adjusting the speed of the filter belt, the pressure and sludge consistency, and inflow rate. This type of press is very useful for both organic and inorganic sludges as well as for mixed sludges. However, the sludge usually has to be chemically conditioned before dewatering. Fig. 4 shows the flow-sheet of a belt filter press.

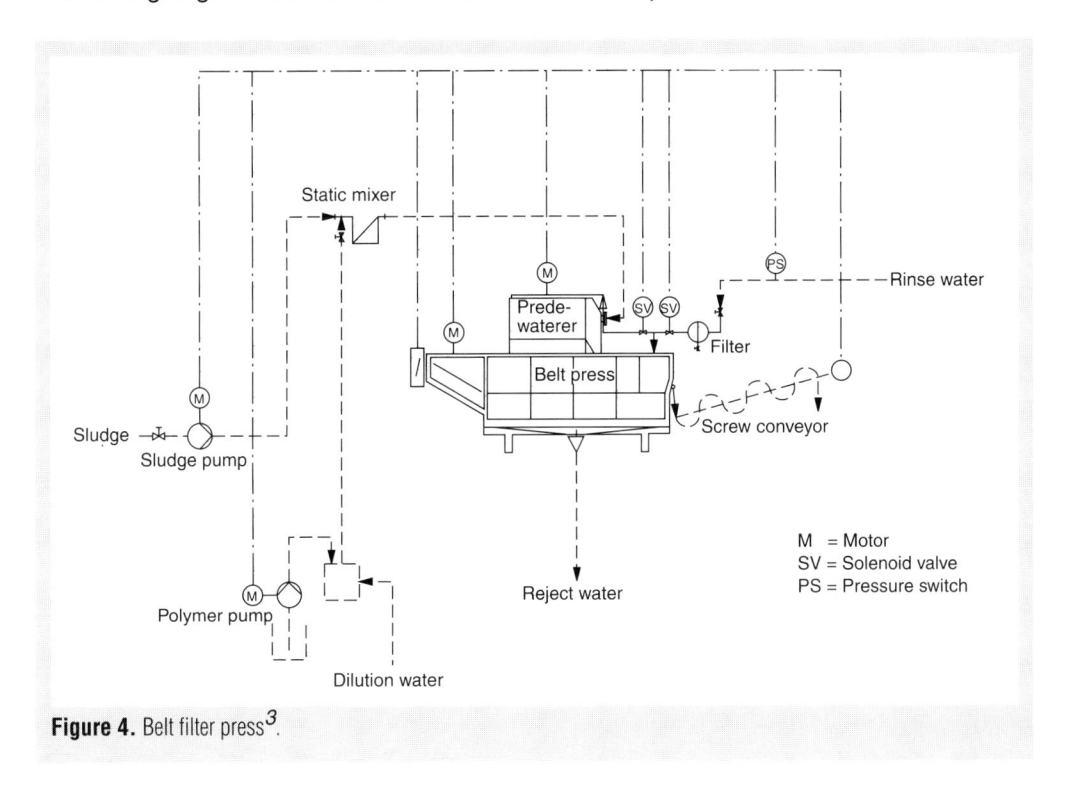

Figure 4. Belt filter press[3].

CHAPTER 8

Screw press
In a screw press, sludge is pressed by a screw conveyor against the inside of a perforated shell, causing water to exit via the holes in the shell. Several different models of screw press have been developed, the basic differences being in the design of the compression zone. Pressure is created by employing a conical screw or a conical shell, or else by installing plates at the sludge discharge end. Fig. 5 shows the design of a Tasster screw press.

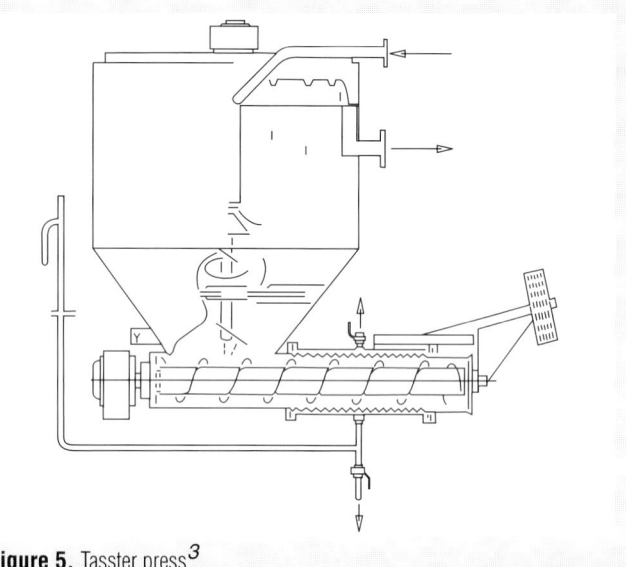

Figure 5. Tasster press[3].

Post-pressing
After preliminary treatment, the sludge is fed between two slowly moving mesh belts under high compression. The pressure on the mesh belts is exerted by water-filled cushions, which press on the belts via grooved rubber blankets.

Heat treatment refers to sludge conditioning (improving water release properties) by means of heat, pressure, and chemicals. The idea is not to oxidize organic matter. The principal reaction taking place in heat treatment is the hydrolysis of lipids, proteins, and carbohydrates, which break the bonds between these compounds and molecules of water. The result is release of intercellular water.

The elevated temperature and pressure required are produced by means of a pump in a pressurized reactor. The best-known heat treatment in current use is the Zimpro process, which can also be used for the wet oxidation of sludge. Sludge is pumped along with air via heat exchangers into a pressurized reactor, where the reactions take place at 170°C–205°C and 10–20 bar pressure. The reaction time is 30–45 minutes. The sludge-water slurry leaving the reactor goes via a heat exchanger to a set of aerators.

In Finland, heat treatment has been tested on both laboratory and pilot scales. According to some results, heating biosludge under pressure at 100°C results in a 10%–20% higher dry solids content for a mixture of biosludge and primary sludge after the belt filter press than for an untreated sludge mixture.

Heat treatment is intended to improve the dewatering properties of difficult sludges. In the wood processing industry, heat treatment would be used mainly for sludges from biological effluent treatment.

A heat treatment plant (Zimpro process) is used in the wood processing industry in Wisconsin, U.S. The plant handles 30–50 metric tons/day of flotation-thickened biosludge from anactivated sludge plant. The sludge consistency is 3%–5%, the temperature is around 200°C, and the pressure is 26 bar. The sludge is then allowed to settle to a solids content of 16%–17%. The gas from the reactor is taken to a packed column scrubber and then back into the process.

Wet oxidation

Complete wet oxidation refers to the (almost) complete destruction of organic matter present in sludge. The elevated temperature and pressure needed can be produced by means of a pump in a pressurized reactor or using the weight of a column of liquid positioned in a deep borehole.

At least four wet oxidation plants based on the Zimpro process are currently in operation in the wood processing industry. In these, sludge and air are pumped through heat exchangers into a pressurized reactor, where oxidation takes place at 200°C–300°C and 120–150 bar. The water leaving the reactor goes via a heat exchanger to a separation unit where solids and gas are removed.

In the vertical reactor vessel (VRV) method, the reactor consists of a deep hole bored in the ground and into which are positioned concentric shafts. This arrangement enables liquid to be circulated between the surface and the bottom. The sludge, which has been treated by crushing or sieving, is fed into the inner shaft. Air or pure oxygen is fed into the sludge at a certain depth.

Wet oxidation breaks down organic material in the sludge. It is particularly useful for difficult biosludges. However, the process is only considered economically feasible at high capacities (> 10 metric tons/day).

In the wood processing industry wet oxidation is used to:

- Break down sludges from effluent treatment
- Recover china clay from sludges.

Its advantages are:

- Considerable reduction in sludge volume
- Resulting solids easy to dry
- Heat recovery option
- Low cost of raising pressure (VRV).

Its disadvantages are:

- Serious corrosion problems are common
- Need to treat the water removed
- High capital costs
- Needs much supervision and management.

CHAPTER 8

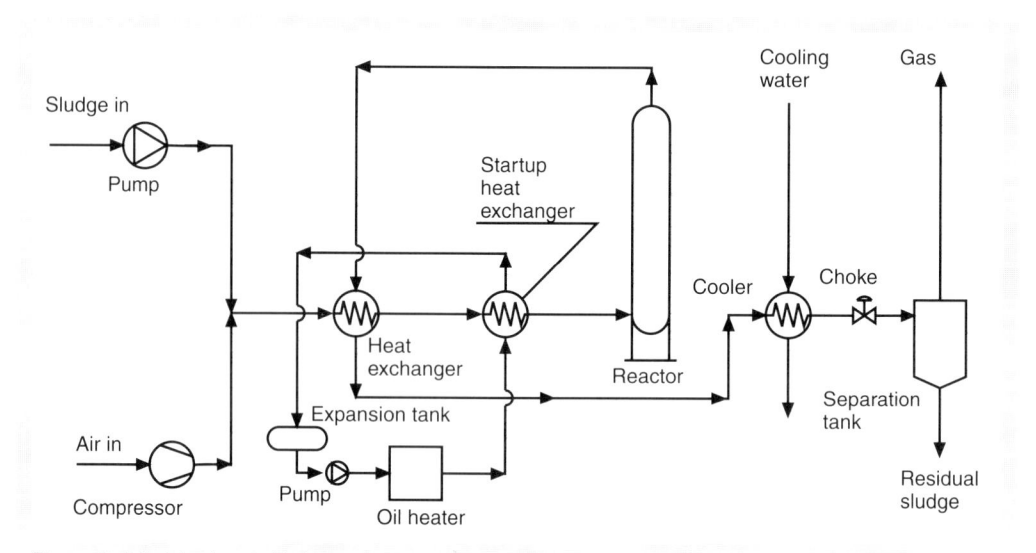

Figure 6. Principle of wet oxidation.

The first wet oxidation plant to be introduced by the wood processing industry (Zimpro) started up in Switzerland in 1978. The plant has a capacity of 13 metric tons/day, 75% of which is primary sludge and the rest activated sludge. The sludge contains 67% china clay. Some 90% of the sludge is burned, and the china clay is recovered in a separate stage. A process like this is in use at a paper mill in the United States.

The first full-scale vertical reaction vessel system was tested with a city treatment plant sludge at Longmont City, CO., U.S. The tests, which were conducted under various conditions in 1984–85, showed that temperature plays a major role in COD reduction. The reductions achieved were 75%–80% for COD, an average of 78% for suspended solids, and an average of 97% for ignition residue.

A wet oxidation plant designed to deal with sludge from treatment plants and slurry from pig farming was started up in 1990 in the Netherlands. The plant has a capacity of 70 metric tons/day and the reactor has a depth of 1 200 m. The water leaving the reactor undergoes separate biological treatment for nitrogen removal. The gas released is taken for catalytic oxidation.

In composting processes, micro-organisms (principally bacteria) break down the organic matter in the sludge under aerobic conditions. Composting can be performed either mechanically or in ridge stacks, the product in both cases being stable humus.

Composting consists of these stages:

- Mixing of sludge and supporting materials
- Aeration
- Storage
- Use.

Composting is suitable for both biosludge and sludge from the primary clarifier. However, even after drying, sludges are seldom suitable for composting as such because they tend to become too compact, thus preventing the proper release of gas. To overcome this, the supporting materials have to be mixed in to provide space for the gas released during decomposition. Frequently used supporting materials include bark, wood chips, and similar biodegradable material. Nondegradable materials such as plastic may also be used, but these have to be screened out of the product and reused.

Several mechanical composting methods are available. All, however, require that the material to be composted have the right temperature and moisture content and that adequate oxygen be available.

The composition of the sludge's organic component, the availability of oxygen and ease of gas release are vital to all composting processes. The moisture content of the material being composted should be not less than about 30% and not more than 60%–75%. The optimum temperature for stabilization is around 50°C–60°C. The pH should be neutral or very slightly acidic. The sludge and supporting material should have a C:N ratio of 25–50:1. In the case of sludge, nitrogen supplementation is usually necessary, often in the form of urea addition. The phosphorus content of sludge is normally adequate. The C:P ratio should be 100:1.

Mechanical composting methods include:

> The **Dano biostabilizer**, which consists of a long drum rotating slowly at approximately 6–60 rpm. Air is fed in through nozzles mounted along the entire length of the drum. Certain nozzles can be used to introduce water in the event that the material being composted is too dry. The temperature reaches about 60°C–65°C at the mid-point of the drum, after which if falls toward the discharge end of the drum. The sludge mixture passes through the drum in 3–6 days. The composted sludge is screened and stored for later use.

> A **BAV reactor** is used to compost sludge mixed with sawdust, bark, or some similar material. In this process, some final product is mixed with the incoming sludge before aeration. The main components are a mixing unit and an aeration reactor. Sludge, final product, and additives are first mixed together. The mixture is fed into the top of the aeration reactor, while air is blown in from below. The final product is stored, ready for use.

> In **layered composting**, a mixture of sludges is led into the top of the composter, from where it is brought downward layer by layer. Air is blown into the composter from below.

> Compost is removed from the bottom and stored, ready for use.

> In a **ridge stack composter**, sludge mixed with some other material such as bark is spread in layers to form ridged stacks. The ridges are 3–4 m high and 3–6 m wide at the base. The cross-sectional shape of the stack is that of half a parallelogram. The material is mixed at intervals of several weeks using a tractor-mounted shovel or some specially constructed implement. The temperature at the center of the stack can reach around 55°C–70°C.

CHAPTER 8

Figure 7. Principle of ridge stack composter.

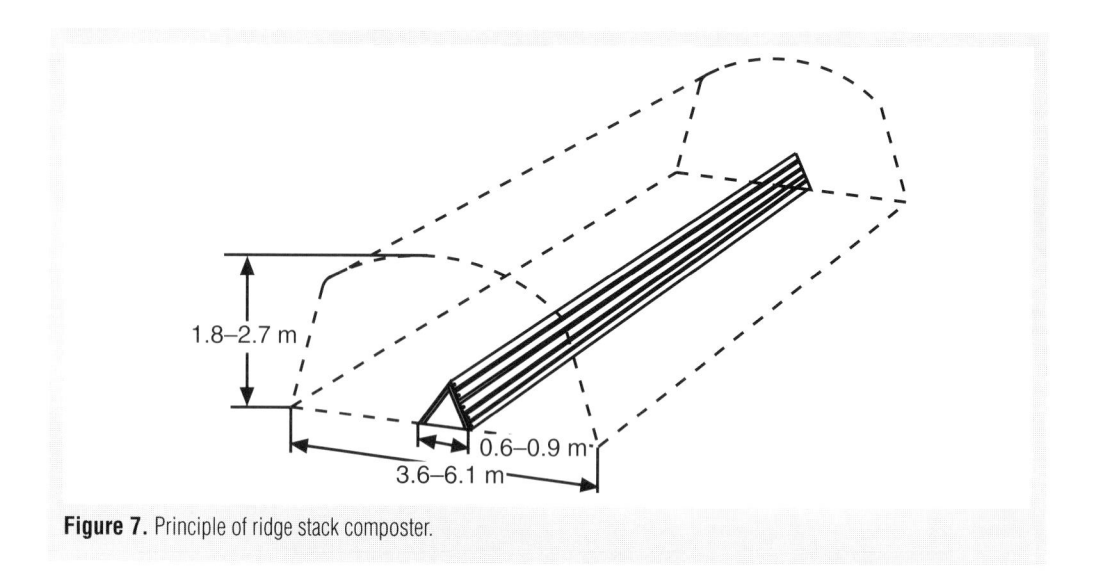

Ridge stack composting can be made more efficient by means of suction (Beltsville method) or by blowing air through pipes into the middle of the stack. Both reduce the need for physical mixing and speed up the decomposition process.

In this composting method, the temperature is generally about 55°C but can be as high as 70°C. Less space is required for composting than with other methods.

The resulting compost is milled and screened ready for use.

Surface water and leachate draining from composting areas must be collected and treated. The volume and composition of such waters depends on the composting method used. Ridge stack composting can cause unpleasant smells if conditions within the stack become anaerobic. Odor problems are less likely with composting systems employing air suction or blowing. The composts produced consist mainly of humus and can be used for purposes such as soil improvement.

One of the advantages of composting in the case of biosludge is that it allows this difficult material to be put to good use, e.g., in soil improvement. The compost is suitable for coarse soils as it retains both moisture and nutrients. It is also an excellent substitute for peat litter.

Disadvantages of composting include the considerable space and labor requirements and the problem of utilizing such large volumes of material.

In Finland, composting is used only to a limited extent for dealing with forest industry sludges. The main method is ridge stack composting employing a tractor-mounted shovel for mixing. The compost produced is used mainly for soil improvement, horticulture, and landscaping of landfill sites. Composting is unlikely to provide the answer to the problem of final placement of forest industry sludges in the near future.

3 Waste incineration

A common way of dealing with forest industry waste is to incinerate it. In the case of waste originating from wood, over 90% is burned for energy recovery. Some of the forest industry's problem wastes, such as solvent and oil residues, are also disposed of by burning.

3.1 Combustion of sludge

Sludge is burned largely to reduce its volume and to reduce it to an inorganic form, which is easier and cheaper to landfill. In some cases, sludge can be dried to sufficiently high dry solids contents to make burning an economically viable option.

In almost all cases, forest industry sludges are burned together with solid wastes derived from wood. Boilers designed for burning sludge are becoming more common, particularly for burning sludges from municipal treatment plants.

Sludges produced by the forest industry with potential for burning include fiber sludges from pulp and paper mills, bark sludges, biosludges from activated sludge treatment, and deinking sludges. The sludges are normally combined prior to mechanical dewatering. Their heat values are usually low or even negative. In practice, burning sludges invariably requires the use of an auxiliary fuel.

Sludge handling and feed to the boiler

The sludge to be burned usually has a dry solids content of 20%–40%, which is achieved by mechanical dewatering (screw presses can give up to 50% dry solids). Once at the boiler plant, no handling other than storage and feed systems are required. Nevertheless, mechanical dewatering can be improved if it leads to more stable combustion and thus to lower emissions.

Where sludge is mixed with bark waste, care should be taken to ensure that the final mixture is as homogeneous as possible. A drag conveyor is useful for carrying the fuel mix to the furnace. The fuel should be fed in as uniformly as possible in order to minimize harmful emissions. Feeding sludge separately into the furnace can also render combustion unstable, even though the fuel ratio is easier to manage in this way.

Sludge combustion processes

Sludge is burned mainly in fluidized bed and grate boilers. Burning of biosludges have also been carried out using recovery boilers at pulp mills. Possible disruptions to the combustion process must be considered when planning to burn sludges. Sludge can severely impair combustion and cause fouling of heat surfaces. Also, bark-burning capacity falls and the volume of flue gases increases. As a result, steam generation decreases, and there may also be corrosion problems.

CHAPTER 8

Grate combustion is less efficient than fluidized bed combustion. This is largely because grate combustion is less stable and more difficult to manage, and the combustion conditions are more difficult to control. With grate combustion, sludge must make up no more than 10%–15% of the fuel mixture; otherwise, proper combustion cannot be achieved (Fig. 8).

Fluidized bed combustion results in good heat transmission and intimate mixing of air and fuel. Combustion is, on the whole, better than in grate furnaces, and emissions are therefore lower. In fluidized bed boilers the combustion conditions can be controlled through regular temperature and pressure measurements, which is not the case with grates. Fluidized bed boilers are divided into bubbling and circulating fluidized bed types. The combustion efficiency is about the same (Fig. 9).

Fluidized bed combustion is by far the best of the current methods for burning sludge. The technique makes it possible to achieve the conditions needed for dioxin destruction (>850°C and adequate combustion time).

Of the emissions from fluidized bed boilers, it is dioxins that have recently caused the most debate. Trace amounts of dioxins are released in some cases when chlorine-containing fuels are burned, and this includes sludges from pulp and paper mills.

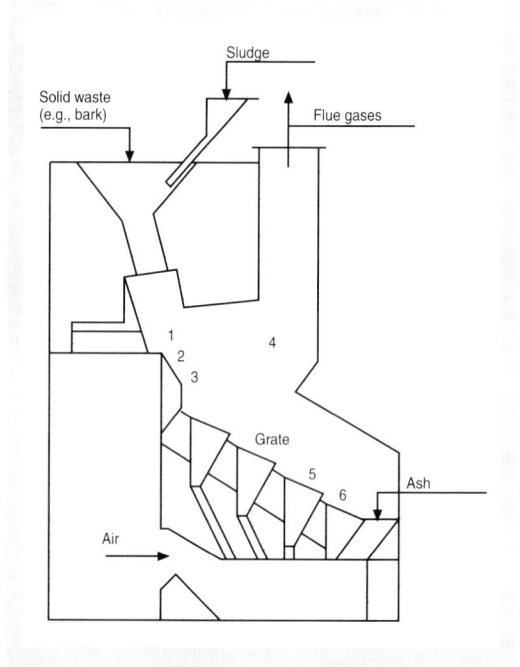

Figure 8. Grate combustion.

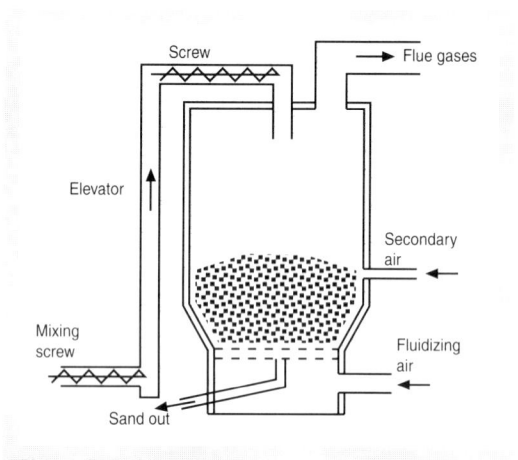

Figure 9. Fluidized bed combustion.

Dioxin emissions from fluidized bed boilers have been measured in Finland, where older plants have released under 1 ng/m^3 (under normal conditions). A limit of 0.1 ng/m^3 has been set in continental Europe, but has not always been reached. Studies show that dioxin emissions can be reduced by adding lime, sulfur, or ammonia to the fuel. It is also important for dioxin removal that combustion be as complete as possible and that the process be properly controlled. The furnace must also be properly cleaned of soot. Dust has to be removed while the flue gases are still hot, and the electrostatic precipitator must be operated at 200°C or less.

Other emissions from the burning of sludges arise in the same way as those from other fuels, and the same methods can be used for their reduction. Emissions of HCl and heavy metals can be reduced in the same way as with dioxins.

The burning of sludge involves both investment and operating costs, even if a boiler plant is already available. These costs are usually higher than the value of the heat recovered. Alternative methods of disposal such as landfill, anaerobic decomposition, or composting therefore must be considered in terms of cost. Combustion is generally a viable alternative when considering how to dispose of increasing volumes of sludge.

4 Landfilling of wastes

Landfilling is one way of disposing of effluent sludges from the forest industry. In 1987, the Finnish forest industry employed 91 industrial and 430 municipal landfill sites for disposal of its wastes. In that year, the forest industry dumped around 730 000 metric tons of various wastes in these landfill sites. Roughly 500 000 metric tons of this was taken to industrial landfill sites and 200 000 metric tons to communal sites.

Forest industry wastes taken for landfill comprise largely sludges, ash, bark waste, and various process wastes. Sludges pose the biggest problems in relation to landfilling; these will be discussed more in the next section.

5 Landfilling of sludges

Before landfilling sludge, it is first dewatered by any of the currently used methods. The aim should always be to achieve the highest possible dry solids content before landfilling.

Factors to be considered when planning and building a landfill site include:

- Environmental suitability of the area in question
- Geology of the area
- Opportunity for inspection of groundwaters
- Environmental impact of runoff water from the site
- Equipment and other installations needed
- Factors related to continuous use of the landfill site
- Composition and volume of sludge
- System used for filling the site.

Before engineering a landfill site, all precautions must be taken to ensure that the site will not cause harm or danger to the surrounding environment. Fig. 10 illustrates a managed landfill site.

CHAPTER 8

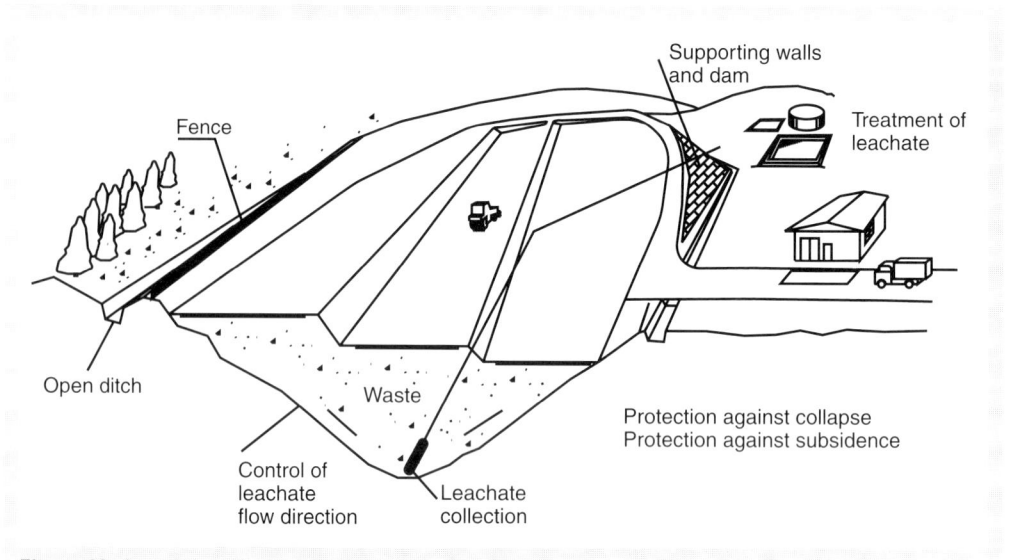

Figure 10. Section through a managed landfill site.

There are several ways of depositing layers of waste in the site. Layers are usually 1–1.5 m in depth and are covered with a thin layer of gravel or earth. A new layer of sludge should only be added after the previous layer has compacted and the water contained in it has filtered downward. When the site is full, it should be landscaped over and any subsequent use properly provided for.

All leachate from the sludge, as well as rainwater and runoff water from the site should be collected and taken to an appropriate treatment plant. The volume of water leaching from a landfill site is normally fairly small, although such water may be highly concentrated. Runoff water, on the other hand, can constitute large volumes, although it is very dilute provided the site has been correctly and quickly filled and capped.

There is a risk of pollution from runoff water and leachate and, for this reason, it is vital to give considerable attention to the design and engineering of landfill sites. Runoff waters and leachates have been found to contain

- 10–40 g/l COD
- c. 8 g/l BOD
- Heavy metals:
- Chromium (0.1–0.5 mg/l)
- Cadmium (<0.05 mg/l)
- Lead (<2 mg/l).

Provision should be made nearby for emergency siting of waste in the event that heavy or prolonged rain temporarily prevents the main site from being used.

Proper management and supervision of a landfill site requires monitoring of the area's groundwater via specially positioned pipes or a system of wells. A constant check should also be kept on surface waters in the area in case of leaks. A proper monitoring program should be drawn up and implemented throughout the year.

Landfilling has been widely used as a means of final disposal of sludges. This option, as well as other methods of disposal, should continue to be available in the future. In some instances, forest industry sludges have been taken along with other waste to municipal landfill sites. However, this is not recommended in view of the large volumes of the sludges concerned and because sludge can interfere with the working of the landfill site.

Runoff water and leachate from landfill sites are collected in basins prior to treatment. Site monitoring includes taking samples from the basins for analysis. During design of a landfill site, provision should be made to collect the landfill gases given off. These can then either be burned or released into the air, provided there is no odor or other problem. If they are properly managed and used, landfill sites do not normally give rise to other emissions.

One disadvantage of the landfill disposal of sludges is the large space required, which means high investment costs when a site is established. Sludges cannot usually be piled very high because of their high water content. Layers of sludge may also cause smell or attract insects to the area.

One Finnish paper mill has the practice of spreading sludges from activated sludge treatment over a specially constructed landfill area. The sludges in question are a mixture of primary and excess sludge dewatered in a belt filter press to around 20%–30% dry solids. As the layers of sludge will not support vehicles, the sludge is formed into closely adjoining ridges. To obtain thicker layers, from time to time sludge is lifted by loader on top of previous layers. This can give layers up to 3–5 m thick.

Some mills have taken their sludge to municipal landfill sites, where it has been spread either separately or together with domestic waste. In some places, both sludge and bark have been composted at landfill sites either as part of final landscaping or for other horticultural purposes.

6 Disposal of other wastes

6.1 Soil improvement

One use for sludges and ash from the forest industry is as soil improvement agents. This is, in fact, becoming more widespread, especially in the United States. In Finland, around 60 000 metric tons of ash were used for soil improvement in 1992.

The use of sludge and ash for soil improvement is limited mainly to agricultural land and forest; some is used in makeup soils. Forest industry sludges are suitable for these purposes because they contain far less heavy metals than, say, municipal sludges. They contain no pathogenic organisms, and have high organic contents. One drawback is the fact that the organic compounds present in forest industry sludges have not been completely identified. The organic chlorine compounds present in sludges from

CHAPTER 8

bleached sulfate pulp production, together with their environmental impact, are one group needing further investigation.

Neither is much known about the effects of chlorine compounds present in ash when used for soil improvement. Today, however, local government authorities in Finland often require ash to be analysed for chlorine compounds before being used for this purpose.

Applied to the land, sludge improves soil quality by preventing the leaching of nutrients and providing added moisture. It also acts as a fertilizer by supplying plants with nitrogen, phosphorus, and trace elements.

The amount of sludge spread per unit area depends on the characteristics of the site and the composition of the sludge. Local regulations also have to be observed. In the United States, for example, sludge has usually been applied at the rate of 50–150 metric tons/ha (calculated as dry matter). A report by Finland's Ministry of the Environment working group dealing with treatment plant sludges recommends a maximum annual application of 1 metric ton of sludge (dry weight) per hectare of agricultural land. As there are no specific guidelines for forest industry sludges, the instructions for using sludge from municipal waste treatment plants should be followed, as applicable, when sludge is used for soil improvement.

Sludge can be used for soil improvement in liquid form straight from settling tanks, after thickening, or after dewatering. There are several ways of spreading sludge on the land, depending on what form the sludge is in. Dewatered sludge is spread in the same way as manure, while liquid sludge is spread using special implements or is injected beneath the topsoil.

The soluble components released from sludges used for soil improvement still require further investigation. Their potential adverse effects depend on factors such as the method of application, monitoring, and adequate supervision.

Use of sludge for soil improvement has the following advantages:

- Utilization of nutrients and trace elements
- Strengthens humus layer and raises its moisture content.

Disadvantages include:

- The need for large areas
- High transport and handling costs
- Possible toxic and other effects incompletely understood
- The fact that sludge cannot be spread throughout the year means that storage facilities are needed; this means higher handling costs and greater manpower requirement.

In Finland, tests have been carried out on the use for soil improvement of sludge from the activated sludge treatment plant at one pulp and paper mill. No final report has been published on the findings.

Sludges have been used for soil improvement in the United States, where some 30 test projects have been carried out. Sludge has been applied in liquid form and also after dewatering. Tests have been performed with all types of sludge, including biosludge, primary sludge, and their mixtures.

Use of sludge for soil improvement is fairly new, as most tests and projects were started in the late 1970s and early 1980s.

6.2 Returning sludge to the production process

While it is often technically possible to return sludges to the production process, the effect may be impairment of product quality and process runnability. The use of biosludge and fiber sludge as raw materials in production processes must always be considered on a mill-by-mill basis.

In some cases, fiber sludge from the primary clarifier is cleaned up and used in paper and board production.

Biosludge and mixed sludges have not been used as raw materials in production processes in Finland, and there are only a few examples worldwide. The main applications studied and tested are:

- Its use as an additive in paperboard and fiberboard production
- Its use in combination with black liquor or spent cooking liquor before evaporation and burning.

6.3 Production of animal feeds from sludge

Activated sludge from municipal treatment plants and from the food industry has been used to produce animal feed, as described in the published results. Sludges from activated sludge treatment of forest industry effluents have also been tested for this purpose.

The Waterloo process is one method for fermenting various wastes to produce protein. Trials have been carried out with materials such as straw, animal manure, and sawdust. Effluent sludge from chemical pulp production has also been tested. The process itself consists of three stages, in which the active organism is the fungus Chaetomium cellulolyticum.

The Waterloo process comprises these stages:

- Thermal and/or chemical treatment (hot water, NaOH or NH_3)
- Aerobic fermentation
- Isolation of the desired product.

A few processes, such as the Carver-Greenfield process, dry biosludge by evaporation to produce an animal feed ingredient.

The biggest problem with these processes is the difficulty of marketing the product.

CHAPTER 8

6.4 Other products from sludge

Other ways of processing and utilizing effluent sludges from forest industry plants include the following, some of which are still being studied:

- As an additive in brick-making
- For road and landfill site constructions
- As a filler in the manufacture of plastics and plywood adhesives
- In the production of ethanol
- As an additive in composting.

References

1. Hynninen, P., "Environmental protection in the forest industry," Lecture sheets from Helsinki University of Technology, Puu-23. 190, Espoo, 1997.
2. Oy Slamex Ab.
3. Tamflow Oy.

CHAPTER 9

Other environmental impacts and their reduction

1	Wood procurement	133
2	Noise abatement	133
3	Transport	134

CHAPTER 9

Other environmental impacts and their reduction

1 Wood procurement

A lot of mistakes have been made in forestry work, particularly that carried out under warm and wet conditions. Some of these mistakes are very difficult to rectify. In certain areas, including the Nordic countries, forest clearance has caused changes to microclimates and local water resources, with deleterious effects on subsequent forest regeneration. Other adverse effects on the natural state of the environment and its multiple use have stemmed from bog drainage and soil preparation carried out with a view to speeding up forest regeneration. These measures have caused substantial increases in loadings to waterways.

This much debated issue has resulted in more environment-friendly forestry methods being introduced or already put into practice. Examples include reducing the size of clear-felling areas, an end to unnecessary drainage, new soil preparation guidelines, and new choices of tree species. Greater consideration for the environment during planning has a major role to play here.

Countries such as Finland, where forest resources are vital to the economy, will probably continue to utilize wood very efficiently, and this will continue to create conflicts. Progress in harvesting techniques will play a key role in this respect. Another issue, and one that remains largely unsolved, is how best to utilize the large amount of smallwood that exists in forests due to their age structure. This is a fundamental question and one with environmental implications.

2 Noise abatement

The regulations pertaining to noise abatement concern noise control in the workplace and control of noise emissions to the surrounding area.

The noise control methods used in industrial plants are:

- Vibration damping and enclosure of individual machines
- Damping of noise arising from flows of intake and exhaust gases
- Sound damping between different parts of the plant using cladding and various structures
- Sound-proofing of individual working areas (control rooms, etc.).

CHAPTER 9

Wherever possible, plans for noise abatement should include measurement of source noise levels and frequencies. Some sophisticated calculation methods are available for this purpose.

3 Transport

As far as transport is concerned, particular emphasis has been placed on minimizing emissions due to spillage of chemicals both on mill sites and elsewhere. This has resulted in the introduction of stipulations regarding the design of transport vehicles and the points of loading and discharge. In the forest industry, the most potentially hazardous chemicals that have to be transported to mills are chlorine, sodium hydroxide, and various acids. Special arrangements have also been made with regard to numerous other chemicals and fuels.

Some products are carried both by road and by ship.

Raw materials (principally wood) have to be transported to the mills, while products leaving the mills also require transport. The forest industry accounts for a high proportion of Finland's total transport of goods, and all stages of the transport chain have some effect on the environment.

Transport on this scale naturally results in loadings to the environment. Some of the impacts are spread over a wide area and represent part of transport loadings in general, while others concern particular mills or their immediate surroundings.

By making special arrangements and using safety equipment, the mills ensure the safe loading and storage of chemicals. Transport planning and routing with a view to minimizing adverse effects is coordinated together with other parties responsible for transport in the same area. A framework exists for action together with local rescue services in the event of any accidents.

CHAPTER 10

Tools for environmental management

1 Economic calculations 137
2 Life cycle analysis 138
3 Environmental systems 139
4 Environmental impact assessment 140
  Bibliography.. 141

CHAPTER 10

Tools for environmental management

1 Economic calculations

The costs associated with environmental protection are monitored both in Finland and elsewhere. In the United States, the body responsible for this monitoring is the Bureau of Census; and, in Germany, it is the Ministry of the Environment. In Finland, the job is handled largely by Statistics Finland.

The methods used for these calculations differ from one country to another, sometimes to greater and sometimes to lesser degrees. The differences are due to the fact that different countries compile their statistics in different ways. There are also differences of opinion as to how environmental protection costs should be defined. Basically, the problem is one of deciding what measures constitute environmental protection. Once made, such definitions must be reviewed from time to time.

In some cases, measures giving rise to environmental protection costs are restricted to those that are not economically viable from a business viewpoint.

The OECD has sought to establish comparability in environmental protection statistics and, to this end, in the late 1970s published recommendations with which most of its member states have tried to conform. Technical specifications were offered as a basis for distinguishing environmental protection costs from other costs. Under this system, a percentage breakdown of costs is not acceptable.

All investment costs are included, namely:

- Purchase of machinery and equipment

- Purchase of buildings and structures

- Value of land purchased or already owned

- Cost of modifications to existing machinery, equipment, buildings and structures

- Production losses due to environmental protection measures.

Operating costs include labor costs, energy, supplies, repairs, servicing, and possibly also rents or leases.

In Finland, the National Board of Waters and the Environment – as well as Statistics Finland, industrial organizations, and the Ministry of the Environment – collect statistics on environmental protection expenditure. Directions relating to the environmental technology to be employed by the pulp and paper industry were produced in the late 1980s as part of a project coordinated by SITRA (the Finnish National Fund for

CHAPTER 10

Research and Development). The National Board of Waters and the Environment, in particular, has since sought to follow these directions. From time to time, the results have been reviewed in connection with the financial planning of environmental protection measures and subsequently entered in committee reports. A Finnish technology program during the early 1990s also calculated expenditure on environmental protection.

In the early 1990s, Statistics Finland also produced a computation model based on EU recommendations. The fact that one set of figures, particularly those relating to costs, is often difficult to compare with another should always be borne in mind, and the bases for calculations should be carefully examined.

2 Life cycle analysis

Life cycle analysis (see Appendix 3) was probably first introduced by the packaging industry as a means of assessing the "environmental impact" of a product and its manufacture. LCA gained considerably in popularity during the 1980s. It involves calculating emissions to the environment resulting from the procurement and consumption of raw materials, energy generation, and the manufacture, use, and final disposal of a particular product. Emissions are calculated per product unit (e.g., per metric ton), and each emission is given a value for use in calculations. The final assessment is thus based on environmental impacts.

LCA permits, at least in theory, a comparison of the environmental loadings and other environmental effects of different products and of the same product manufactured in different ways. It is now becoming considerably easier to make the appropriate choices and indeed to manage the analysis as a whole.

Numerous models have been developed for life cycle analysis, and the number is still growing. Table 1 lists some of those currently in use.

The models are designed to take full account of the situation as a whole and also to reflect accurately observations made in the field. The various models differ, sometimes considerably, from each other.

Life cycle analysis is perhaps most useful in product planning, in which alternative products are compared in terms of their likely environmental effects. However, the choices made reflect the weaknesses of LCA: despite the values allocated to different emissions, assessments based on environmental impact are in practice only approximate. The effects of the different emissions also vary according to location: for example, the same level of the same emission can have different effects depending on the nature of the receiving water.

LCA may also be used as a guide to the choice of packaging methods and in audits made preparatory to applying for the right to use an environmental label (see Appendix 4). The use of LCA is likely to continue to grow considerably and to affect production as a result of market pressures. Here, the standardization organizations have identified a niche market, and standards for testing a variety of products and types of product are under preparation. The "standardization of standards" is also a question of current importance.

Table 1. Some computer-based models for life cycle analysis.

MODEL	DEVELOPER/COUNTRY	SUBJECT
1. I. Boustead	INCPEN ym./England	packings
2. Eco-model	Frauhofer-Institution/Germany	packings
3. EPS	Frauhofer-Institution/Germany	wide, estimation pocedure
4. GBA	University of Stutgart/Germany	all products, preliminary
5. GEMIS/TEMIS	Öko-Institution ja GH Kassel/Germany	wide
6. IDEA dBASE IV	HASA/Austria(Finland)	wide
7. LCA Inventory Tool	Chalmers Industry-Technics/Sweden	
8. Ökobilanz	EMPA/BUS/BUWAL/Switzerland	wide, estimation procedure
9. Oekopack	Lss Lahyani Software Solutions/Switzerland	packings/based on BUWAL-procedure
10. REPA	Franklin Associates Ltd/USA	
11. Simapro	CML (PRe addings)/Netherland	wide
12. G. Sundström	Sundström Miljöbalans AB/Sweden	packings
13. Umwelt controlling	IÖW,PSI/Germany	

3 Environmental systems

The ISO 9000 series of quality system standards has gained rapid acceptance as a route to securing the high quality of operations. A number of different environmental protection "standards" have been prepared based, to varying extents, on quality systems. Industry has started to introduce systems based on ISO 14001 and British Standards at certain mills (see also Appendix 2). The idea is to establish environmental protection-related guidelines suitable for all levels and all mills. At many mills, the work will probably be organized in such a way that the job of keeping the guidelines up-to-date and monitoring the success of measures taken falls, at least partly, to the person responsible for the performance of equipment designed to protect the environment. Establishing a suitable system requires a full review of the goals set for environmental protection and of the facilities available, plus a listing of all observed deficiencies. Instructions on how to act need to be produced for the different people within the organization. These instructions must also be reviewed from time to time and any necessary additions made. The mill may then advertise that it meets the requirements of an "environmental quality system."

Although work done to this end has so far received certification at only a few plants, it nevertheless is rapidly becoming more common. Environmental quality systems have also been criticized for including unnecessary activities, and changes for the better can be expected as more experience is gained.

In many companies, environmental audits not connected with environmental systems are likely to be replaced, at least in part, by environmental quality systems. Environmental audits were intended to achieve broadly the same goals as with environmental quality systems. Another result will be to bring greater consistency and harmony to the environmental obligations and rights accorded to different companies.

4 Environmental impact assessment

A whole series of concepts, including emission parameters, has been devised to assess environmental impacts. Emission parameters, such as the BOD and COD loadings of effluents, have been discussed in previous chapters. Some emissions can be monitored in terms of absolute amounts, examples being sulfur dioxide and hydrogen sulfide. Environmental impact assessment is also concerned with quality recommendations such as those for air quality. Similarly, natural waters can be classified in terms of their suitability for specified purposes. In Finland, this classification system was introduced in the late 1980s following some ten years of preparatory work.

Table 2. Classification of waters for different uses.

Basis for classification
- Health-related factors
- Factors related to odor and taste
- Factors having a temporary effect on water quality
- Variables depicting water quality in general.
Uses of water and watercourses according the classification
- Recreational use
- Suitability as raw water
- Suitability for fish
- General classification.

Quality classifications for both air and water are based on actual impacts. The most important criteria for classification relate to the effects on people's health. Initially, air quality classifications were introduced in Finland by the National Board of Health, whereas today they mainly take the form of rulings by the Council of State. In the case of waters, other factors such as fish stocks and commercial fishing interests are also considered.

Basically, the quality classifications used in environmental impact assessment represent a method of taking into account people's various activities and values. It is probably no exaggeration to say that all changes taking place in the environment have some kind of adverse influence on something. How serious these are depends on who is making the assessment and on the scope of the assessment.

The sciences that play the key roles in assessing environmental impacts are limnology, meteorology, and geology. These are frequently supplemented with knowledge of chemistry, physics, and mathematics. Microbiology also has an important role to play. Indeed, most environmental impact investigations are in practice inter-disciplinary.

Bibliography

1. Hynninen, P., "Environmental protection in the forest industry," Lecture sheets from Helsinki University of Technology, Puu-23. 190, Espoo, 1997.

APPENDIX 1

Appendix 1.

1	**Environmental Legislation, Regulations and Executive Authorities**	**143**
1.1	General	143
1.2	European Environmental Legislation	143
1.3	Finland	150
1.4	Sweden	155
1.5	Norway	159
1.6	Germany	161
1.7	France	165
1.8	United Kingdom	169
1.9	Italy	172
1.10	Spain	173
1.11	Portugal	176
1.12	The United States	177
1.13	Canada	195
1.14	International Conventions	198
	References	206

Appendix 1.

August 1998

1 Environmental Legislation, Regulations and Executive Authorities

1.1 General

Even today, in the late 1990s, environmental legislation still differs considerably from one country to another. Marked variation also occurs in other administrative practices applied in major pulp and paper-producing countries, on the part of norms and regulations, permit procedures, and implementation authorities. The reasons for this are the historical, economic, and social differences between the countries as well as their varied level of industrialization.

The European Union recently made efforts to harmonize environmental legislation with the issue of council directive 96/61/EC concerning Integrated Pollution Prevention Control (IPPC) in September 1996. After a transition period, the implementation of this directive will be gradual in the member countries. Other examples of voluntary harmonization and development in environmental protection are the international ISO 14000 series standards and the respective European Union council regulation (EEC 1836/93) on Eco-Management and Audit Scheme (EMAS).

1.2 European Environmental Legislation

Present State and Principles of European Environmental Law

European environmental law as it now is originates from the entry into force on Nov. 1, 1993, of the Treaty on European Union[1]. The objectives of Community policy are formulated in the first paragraph of Article 130r of the Treaty as follows:

- Preserving, protecting, and improving the quality of the environment
- Protecting human health
- Prudent and rational utilization of natural resources
- Promoting measures at the international level to deal with regional or worldwide environmental problems.

Also, Article 130r[2] sets out the principles on which European environment policy is based:

- The high level of protection principle
- The precautionary principle
- The prevention principle

Appendix 1

- The source principle
- The polluter pays principle
- The integration principle
- The safeguard clause.

For the pulp and paper industry, the source and polluter pays principles are particularly important. According to the source principle, environmental damage should preferably be prevented at source, rather than by using "end-of-pipe technology." The polluter pays principle means that charging polluters the cost of action to combat the pollution they cause will encourage them to reduce that pollution and endeavor to find less polluting products or technologies.

Enforcement of Community environmental law is primarily the responsibility of the member states. Generally, the regulations enacted by the Council of the European Union are aimed to be brought into effect in the member countries gradually, after certain transition periods. The member states are requested to inform the Commission of their national standards and regulations so as to bring them in line with EU legislation (Notification Agreement C9/1 and Directives 83/189 and 94/10).

The Environmental Action Programs

The institutions of the EC have produced five action programs in the environmental sector since 1973[1]. The current, fifth, environmental action program, titled "Towards sustainability," was approved by the Council in 1993. The program is strategic and consultative and mainly focuses on the agents and activities which deplete natural resources and otherwise damage the environment, rather than waiting for environmental problems to emerge before they are dealt with. One of the main elements of the program is the establishment of a general consultative forum comprising representatives of enterprise, consumers, unions and professional organizations, nongovernmental organizations, and local and regional authorities. Its main task will be to consult the Commission on any problem relating to the Community's environmental policy.

Regulations for Environmental Impact Assessment

The Council has adopted a directive on EIA (85/337/EEC) which introduces a common system of EIA procedure in the member states[1]. The directive shall apply to public and private projects which are likely to have significant effects on the environment.

The directive distinguishes between projects which are always made subject to an assessment (Article 4/1) and those which are made subject to an assessment only where the member states deem it warranted (Article 4/2). Major projects listed under Article 4/1 include crude-oil refineries, thermal power stations and other combustion plants, nuclear power stations, the cast-iron and steel industry, waste disposal installations, and various public infrastructure investments. Article 4/2 comprises projects such as certain industrial installations for the production of electricity, parts of the chemical and food industry, and the textile, leather, wood, and paper industries.

The directive gives the member states considerable freedom to decide the manner in which the assessment should be carried out. At least the following information is required, however :

- A description of the project comprising information on the site, design, and size of the project
- A description of the measures envisaged in order to avoid, reduce, and, if possible, remedy significant adverse effects
- The data required to identify and assess the main effects which the project is likely to have on the environment
- A nontechnical summary of all this information.

Water Pollution Control

Two main types of current regulations can be distinguished:

- Directives setting out quality targets (e.g., for drinking water, bathing water, fishing waters, and shellfish waters)
- Directives relating to the protection of the aquatic environment against pollution by dangerous substances (e.g., "black" and "gray" lists).

The main directives are reviewed briefly below.

Framework Directive (76/464/EEC)

This directive, concerning emissions and discharges to the aquatic environment, applies to inland surface water, territorial waters, internal coastal waters, and ground water. The dangerous substances are divided into two groups:

- "Black list": organohalogen, organophosphorus and organotin compounds, mercury and cadmium and their compounds, carcinogens, persistent mineral oils and hydrocarbons, and certain persistent synthetic compounds. **Pollution by "black list" compounds should be eliminated.**
- "Gray list": certain heavy metals, biocides, cyanides, and fluorides. **Pollution by "gray list" compounds must be reduced.**

Groundwater Directive (80/68/EEC)

This directive is related to the framework directive 76/464/EEC. The directive requires **prevention** of the discharge into groundwater of certain substances (contained in List I in the Annex) and **limitation** of the discharge into groundwater of other substances (contained in List II in the Annex). The two lists of the groundwater directive do not correspond entirely with those of directive 76/464.

The present groundwater directive does not contain any references to methods of analysis or to standards.

Appendix 1

Other Directives

Other directives concerning water pollution control as follows have been issued:

- The Shellfish Life and Growth directive (79/923)
- Fish Life Water directive (78/659)
- The Surface Water for Drinking directive (75/440)
- The Nitrate directive (91/676)
- The Titanium Dioxide directive (78/176).

Conventions concluded by the Community

The Community is a party to a large number of regional and international conventions in the environmental field. The most important are[1]:

- The 1974 Paris Convention for the prevention of marine pollution from land-based sources and the 1986 Paris Protocol amending the Convention
- The 1974 Helsinki Convention on the protection of the environment of the Baltic Sea area (as revised in 1992)
- The 1976 Barcelona Convention for the protection of the Mediterranean Sea against pollution, and various protocols to the Convention
- The 1976 Bonn Agreement on the protection of the Rhine against chemical pollution
- The 1983 Bonn Agreement for co-operation in dealing with pollution of the North Sea by oil and other harmful substances
- The 1990 Lisbon Accord of Co-operation for the protection of the coasts and waters of the Northeast Atlantic against pollution.

In addition, the following conventions have been signed by the Community:

- The 1982 UN Convention on the law of the sea
- The 1992 Helsinki Convention on the protection and use of transboundary watercourses and international lakes.

Air Pollution Control

Community legislation for the prevention of air pollution focuses on the following areas[1]:

- Establishing minimum air quality standards and maximum emission standards for certain harmful substances
- Reducing the emission of pollutants from cars and other motor vehicles
- Protection of the ozone layer.

Air quality directives

During the 1980s, the Council adopted the following directives:

- Directive 80/779 for sulfur dioxide and suspended particulates
- Directive 82/884 for lead and lead compounds
- Directive 85/203 for nitrogen dioxide.

The directives contain limits and guide values for pollutants and provisions on establishing measuring stations, sampling, consultation in the event of transfrontier pollution, and standstill requirements.

Directive 84/360 on air pollution from industrial plants

The framework directive contains general obligations for prevention of air pollution from industrial plants operating in the energy industry, production and processing of metals and minerals, the chemical industry, and waste disposal.

Directives limiting emissions from industrial plants

Emission limit values are specified in the following directives:

- Directive 88/609 on the limitation of emissions of certain pollutants into the air from large combustion plants
- Directives 89/369 and 89/429 on the prevention of air pollution from new and existing municipal waste incineration plants
- Directive 94/67 on the incineration of hazardous waste.

Polluting substances for which limits have been set include:

- Sulfur dioxide and other sulfur compounds
- Nitrogen oxides and other nitrogen compounds
- Carbon monoxide
- Hydrocarbons
- Heavy metals and their compounds
- Dust, asbestos, glass, and mineral fibers
- Chlorine and its compounds
- Fluorine and its compounds.

Protection of the ozone layer

The current Community law is laid down in Regulation 3093/94. In this regulation, detailed provisions are introduced on the production, supply, and use in the Community

Appendix 1

of certain ozone-depleting substances (e.g., chlorofluorocarbons or halons). A directive on air pollution by ozone has been issued (92/72) to harmonize procedures for:

- Monitoring
- Exchanging information
- Informing and alerting the population about air pollution by ozone.

Other measures to prevent air pollution

Sulfur content of certain liquid fuels is regulated in directive 75/16. After Oct. 1, 1994, the following regulations are in force:

- The member states must prohibit the marketing of diesel fuel oils if their sulfur compound content exceeds 0.2 percent by weight. After Oct. 1, 1996, the respective content is 0.05 percent. The sulfur content of other gas oils is also lowered to 0.2 percent (except aviation kerosene).

To limit carbon dioxide emissions, directive 93/76 has been promulgated. The directive requires the member states to establish and implement programs to limit CO_2 emissions by improving energy efficiency.

Conventions concluded by the Community

The most important conventions, which the Community has signed, are:

- The 1979 Geneva Convention on long-range transboundary air pollution and the Protocols of Geneva (1984: monitoring and evaluation), Helsinki (1985: sulfur compounds) and Sofia (1988: oxides of nitrogen)
- The 1985 Vienna Convention for the protection of the ozone layer and the 1987 Montreal Protocol on substances which deplete the ozone layer
- The 1992 Framework Convention on Climate Change (Rio de Janeiro).

Waste Disposal

In 1989, the Commission set a general strategy and priorities for waste management as follows[1]:

- Prevention or reduction of waste at source as the highest priority
- Promotion of recycling and reuse
- Harmonization of standards for waste disposal by dumping or incineration based on a high level of environmental protection
- Tightening up of existing rules on the transport of waste
- Cleaning up of sites polluted by waste.

Framework directive on waste (75/442)

The framework directive lays down general rules that apply to all categories of waste. The annex to the directive lists 15 specific categories of waste and adds, for safety reasons, a final category referring to "Any materials, substances, or products which are not contained in the above categories."

The framework directive contains several Articles, where special obligations are laid on the member states for waste management, including requirements about waste management plans and provisions on reporting, periodic inspections, and record-keeping by undertaking.

Packaging and waste packaging directive (94/62)

The directive covers all packaging on sale in the Community and all packaging waste, whether it is used or released at the industrial, commercial, office, shop, or any other level. The directive aims at increased recycling of packaging material: e.g., within five years of the implementation date, between 50% and 65% by weight of packaging waste must be recovered, and between 25% and 45% by weight of all the material contained in packaging must be recycled.

In addition, the directive requires the member states to provide for:

- The return and/or collection of used packaging and/or packaging waste from the consumer
- The reuse or recovery of packaging and/or packaging waste collected.

Other waste directives

The following directives are currently in force:

- The hazardous waste directive (91/689)
- The waste oils directive (75/439).

Integrated Pollution Prevention and Control (IPPC) Directive (96/61/EC)

On Sept. 24, 1996, the Council adopted the IPPC directive. This directive is of great importance in harmonizing environmental legislation within the European Union. The purpose of the directive is crystallized in Article 1, which:

- Aims to achieve a high level of protection of the environment by laying down measures designed to prevent or, where that is not practicable, to reduce emissions in air, in water, and on land from the activities specified in Annex I of the directive, including measures concerning waste.

The IPPC directive will be applied without prejudice to EIA directive 85/337/EEC. The most important principles and obligations included in the directive are:

- Permits for new and existing installations (Articles 4, 5)
- Requirements of applications for permits (Article 6)

Appendix 1

- Decisions and conditions of permits (Articles 8, 9)
- Best available techniques (BAT), environmental quality standards and developments in BAT (Articles 10, 11)
- Exchange of information (Article 16)
- Community emission limit values (Article 18).

Categories of industrial activities which are required to carry out IPPC procedure are listed in detail in Annex I. Main industry sectors are the following:

- Energy industries
- Metals production and processing
- Minerals industry
- Chemical industry
- Waste management
- Other activities (including, e.g., pulp and paper industry).

Annex II lists the directives in force which are to be applied in IPPC procedure when necessary.

Annex III contains an indicative list of the main polluting substances to be taken into account when relevant for fixing emission limit values.

Annex IV specifies "considerations to be taken into account when determining best available techniques, bearing in mind the likely costs and benefits of a measure and the principles of precaution and prevention."

The member states shall adopt the laws, regulations, and administrative provisions necessary to comply with the IPPC directive no later than three years after its entry into force.

1.3 Finland

Water Pollution Control

A basic law (Act No. 264/61) for water pollution control came into force in 1961. This law promulgated rules for and restrictions on the use of water resources (e.g., water supply, fishery, traffic, generation of hydro power, recreation areas, receiving areas for wastewater release).

The Water Act has been supplemented during the 1970s, 1980s, and 1990s (e.g., by restrictions on the release of toxic pollutants, protection of ground water, standards for potable water). With respect to the pulp and paper industry, the basic law includes rules and regulations for granting a wastewater discharge permit. The Water Act does not include any general standards for pollution loads of pulp and paper industry wastewater. Instead, the Water Act requires constraints to be specified (e.g., on allowable pollution loads, measures to decrease water pollution and monitoring programs to follow the effects of pollutants on the receiving water course) in the operating license of every individual wastewater discharge permit granted by a regional Water Rights Court. Most commonly, restrictions on pollutants such as COD_{Cr}(BOD7), AOX, and phosphorus are imposed.

The basic law allows discharge permits for certain periods only, and a new application must be prepared before the expiration of the existing permit.

The Finnish Government made a decision in October 1988 on a basic program for water pollution control up to 1995 including some restrictions for the pulp and paper industry (COD_{Cr}, AOX, phosphorus). A continuation of this program up to 2005 is under way. For the pulp and paper industry in Finland, the recommendations of the Helsinki Commission (HELCOM) and the Nordic Council of Ministers must also be followed. These recommendations are described in detail later in the section International Conventions.

Air Pollution Control

Parliament approved the basic Air Pollution Control Act in October 1982. After 1982, the issuance of a statutory order implemented the law more precisely. In 1987, restrictions on gaseous sulfur emissions from kraft pulp mills considered gaseous emissions of sulfur dioxide acceptable if they did not exceed 4 kg per ton of pulp for new mills and 6 kg for old mills. Subsequently, a new target for gaseous sulfur emissions from pulping processes of 2 kg SO_2/t for new and rebuilt mills has been introduced .

In April 1996, a new amendment to the Air Pollution Act came into force. The main change in the amendment is the replacement of the air pollution control announcement procedure by a new comprehensive environmental permit procedure that incorporates air pollution control applications.

The Finnish Government issued two decrees on ambient air quality recommendations and restrictions in June 1996 as follows:

- Decree (No. 480) on recommendations on ambient air quality and target levels of sulfur precipitation (See Table 1)

- Decree (No. 481) on ambient air limit values to prevent health damage (See Tables 2 and 3).

Table 1. Ambient air quality recommendations in Finland.

Compound	Value (20 °C, 1 atm)	Statistical definition
Carbon monoxide (CO)	20 mg/m^3	Hourly value
	8 mg/m^3	Moving average of 8 hourly values
Nitrogen dioxide (NO$_2$)	150 µg/m^3	One month's hourly values 99. percentile
	70 µg/m^3	The second highest daily value of one month
Sulfur dioxide (SO$_2$)	250 µg/m^3	One month's hourly values 99. percentile
	80 µg/m^3	The second highest daily value of one month
Particulates, total suspended particulate (TSP)	120 µg/m^3	One year's daily values 98. percentile
	50 µg/m^3	Annual average
Respiratory particulates (PM$_{10}$)	70 µg/m^3	The second highest daily value of one month
Total amount of odorous sulfur compounds (TRS)	10 µg/m^3	The second highest daily value of one month (TRS expressed as sulfur)

Appendix 1

Table 2. Ambient air limit values to prevent health damages in Finland.

Compound	Limit value	Statistical definition (20 °C, 1 atm)
Nitrogen dioxide (NO$_2$)	200 µg/m^3	One year's hourly values 98. percentile
Sulfur dioxide (SO$_2$)	80 µg/m^3	Average of one year's daily values
	250 µg/m^3	One year's daily values 98. percentile
Particulates, total suspended particulate (TSP)	300 µg/m^3	One year's daily values 95. percentile
	150 µg/m^3	Annual average
Lead (Pb)	0.5 µg/m^3	Annual average

Table 3. Limit values for ozone in ambient air in Finland.

Reason	Limit value, µg/m^3 (20 °C, 1 atm)	Statistical definition
Protection of health	110	Moving average of 8 hourly values
Protection of vegetation	200	Hourly value
	65	Daily value
Public information	180	Hourly value
Public warning	360	Hourly value

In decree No. 480, the most important restriction for the pulp industry is the target value for all TRS compounds, 10 µg/m^3. To reach this level in the vicinity of kraft pulp mills, extensive odor control measures must be used.

Sulfur and nitrogen emissions in power generation are also restricted. In this respect, the Finnish Government has issued the following decisions:

- Reducing sulfur dioxide emissions from power stations and power boilers using coal (256/1990): maximum allowable sulfur dioxide release varies between 140 and 230 mg/MJ depending on the age and size of the power boilers

- General guidance on reducing nitrogen oxide release from power boilers and gas turbines (527/1991): maximum allowable nitrogen dioxide release varies between 50 and 150 mg/MJ in new plants and between 80 and 230 mg/MJ depending on the type of fuel and power plant

Particulate emissions standards for power boilers have also been issued by the Government (decision No. 368/1994).

Allowable sulfur contents of fuels have also been given in Government decisions.

- Sulfur content in coal (888/1987): sulfur content in imported coal since 1994 can be max. 390 mg/MJ.

- Sulfur content in heavy oil (453/1992): since 1993, maximum allowable sulfur content by weight is 1 percent.

Waste Disposal

The basic law, Waste Management Act No. 673/78, on solid waste disposal was issued in 1978. Subsequently, a decree (307/79) was enacted which includes claims for hazardous waste collection and handling, establishment of landfills, preventing soil pollution, and solid waste collection and handling.

New amendments to the Waste Management Act came into force in 1993 (1072/1993, 1390/1993). In these supplementary laws, major emphasis is on decreasing and preventing solid waste generation and decreasing the harmful and hazardous effects of solid waste as landfill. As a result of these regulations, the amount of landfill has decreased rapidly during the 1990s and control and operation of landfill has been intensified.

Regulations for Environmental Impact Assessment (EIA)

Environmental impact assessment (EIA) procedure is obligatory in major investment projects based on the law which was promulgated in 1994 (468/1994). Greenfield pulp and paper mills having a daily production over 200 tons are obliged to carry out EIA. The documents required in EIA consist of project evaluation and description, evaluation of possible options, specification of environmental effects, pollution control and monitoring plan, and other specified data and information.

Eco-Management and Audit Scheme (EMAS) Law 1994

The EMAS Law (1412/94) applies the respective European community EMAS regulation (1836/93) in Finland.

Economic Instruments

Economic instruments have mostly been introduced for fiscal purposes. At the end of the 1980s, however, a number of economic instruments were introduced for environmental purposes[2]:

- Air pollution abatement

    The present tax system consists of an environmental tax in addition to the basic excise duty. At the beginning of 1990, Finland introduced the first CO_2 tax in Europe. This environment tax was imposed on fossil fuels (except automotive fuels) based on carbon content. The excise duties on liquid fuels and on certain energy sources were revised in 1993–94, in line with EU regulation.

    Excise duty on liquid fuels is levied on automobile gasoline and diesel oil, as well as on light fuel oil for commercial, industrial, or heating purposes and on heavy oil. Excise duty on certain energy sources is levied on coal, milled peat, natural gas, electricity, and pine oil.

    In 1995, the environmentally-based duty on carbon content was FIM 38.3 per metric ton of carbon dioxide and on energy content, FIM 3.5/MWh.

Appendix 1

- Water pollution control

    Recently, charges on artificial fertilizers combating oil pollution have been introduced.

- Waste disposal

    The Waste Tax Law (495/1996) determines the tax on solid waste to be disposed of by landfill. The tax is FIM 90 per metric ton of waste.

    The waste tax is levied on waste delivered to public landfills and comparable sites set up for receiving waste produced elsewhere. No tax is levied on wastes delivered to private landfills, primarily those set up by industry and production units for their own use. No tax is levied on wastes such as stone material, contaminated soil, composting waste, or deinking waste from paper mills.

- Environmentally motivated subsidies

    The Ministry of Trade and Industry awards grants for developments and investment projects which promote energy efficiency or reduce environmental hazards in energy production.

    The Center for Technological Development provides development loans and grants to companies in order to promote development of new technology.

    Environment conservation loans or environmentally motivated tax concessions may be applied in environmental investments.

Executive Authorities

Until the early 1990s, environmental permit procedures were quite complex as many authorities were involved in approvals of applications: the Ministry of the Environment, County Governments, Regional Water Rights Courts, and municipal authorities.

In 1991, a law on environmental permit procedures (735/1991) came into force. In this law, all permit procedures are centralized under one authority: either a regional environment agency or a municipal environmental authority.

The Ministry of the Environment supervises a total of 13 local environmental agencies. In addition, a supporting expert organization, the Finnish Environment Institute, works in close cooperation with the local environment agencies.

The Ministry of the Environment issued a preliminary statement in 1996 on centralization of environmental legislation according to the EU IPPC directive. According to the statement, the new environmental law would include all present laws and regulations and only one environmental permit would be required. The environmental permit would be appropriated by municipal environmental boards. The new environmental law is to be in force by the end of 1999.

1.4 Sweden

General Environmental Legislation

In Sweden, environmental legislation has been developed since the late 1960s as a complete sector. Major environmental laws in Sweden are:

- The Environment Protection Act (1969)
- The Act on Chemical Products (1986)
- The Ordinance on Hazardous Waste (1986)
- The Natural Resources Act (1987)
- The Environment Protection Ordinance (1989).

The essential aim of the principle laws (1969, 1989) is to prevent or significantly decrease environmental disturbances, caused by, e.g., industrial activities, using technically and economically feasible measures. The acts contain rules defining the conditions upon which a polluting activity may be permitted. The basic principle is that disturbances must be prevented where this is practically feasible and that unnecessary disturbances should not be tolerated under any circumstances.

The above-mentioned general principles are implemented mainly through a comprehensive licensing system. Construction, expansion, or other changes to named factories and plants requires a permit, while notification will suffice for minor pollution sources.

The Act on Chemical Products deals with chemical products and their potential to be harmful to man and the environment.

The Natural Resources Act contains basic regulations on the conservation and development of natural resources and special regulations for certain geographical areas including Government approval of permits for certain classes of industrial development.

Disposal of chemicals is regulated by the Ordinance on Hazardous Waste. It applies to chemicals such as solvents, oil waste, acid and alkaline waste, heavy metals, PCB, and pesticides. Disposal, transportation or exportation of these substances requires special permission.

Water Pollution Control

No general standards or norms are applied at present. In the case of individual discharge permits, the authorities usually introduce two types of allowable discharge values: limit values and guide values. The limit value may not be exceeded. The guide value may be exceeded but, if it is, certain measures must be taken to avoid further excess. Limit and guide values are typically defined as annual or monthly average absolute loads. Water pollutants which are usually required to be specified are COD_{Cr}, BOD_7, AOX, P-tot, and N-tot. Standards for frequency and analysis are shown in Table 4[3].

Appendix 1

Table 4. Effluent monitoring requirements and standars in Sweden[2].

Pollutant	Statistical definition	Analyzing method
COD$_{Cr}$	Hourly value	SS 02 81 42
	Moving average of 8 hourly values	
	One month's hourly values 99. percentile	
	The second highest daily value of one month	
	One month's hourly values 99. percentile	
	The second highest daily value of one month	Filtering according
	One year's daily values 98. percentile	to SS 02 81 38
	Annual average	filter fabric 70 um
	The second highest daily value of one month	
	The second highest daily value of one month (TRS expressed as sulfur)	
	In addition:	
	One weekly sample per week.	
	Analyses of COD of nonsettled	
	and nonfiltered sample	
P-tot	One weekly sample per week	SS 02 81 02
N-tot	One weekly sample per week	SS 02 81 01
BOD$_7$	One monthly sample per month	SS 02 81 43
AOX	One weekly sample per week in bleaching plants using chlorine chemicals	SS 02 81 04
	One weekly sample per quarter in paper production	

In some cases, suspended solids and chlorate loads are also required to be measured and reported.

Air Pollution Control

Gaseous sulfur release from the pulp industry has been restricted since the late 1980s. Maximum allowable annual gaseous sulfur release for kraft mills was set at 1.5 kgS/metric ton and for sulfite mills 3 kgS/metric ton. In addition, major emphasis in air pollution control is laid on reduction of odor compounds, particulates, and lately nitrogen oxides as well.

On the basis of the environmental discharge permits granted, typical current restrictions for kraft pulp mills are:

- Max. annual average gaseous sulfur emission, 0.5 kgS/metric ton
- Min. availability of odor control systems, 99%
- Max. hydrogen sulfide concentration in recovery boiler flue gases, 10 mg/m^3 normal dry gases (max. 5% running period/month)

- Max. hydrogen sulfide concentration in lime kiln flue gases, 50 mg/m³ normal dry gases (under max. 10% running period/month)
- Max. annual average NO_2 release from bark boiler, 100 mg/MJ
- Max. allowable particulate emissions in recovery boiler and lime kiln flue gases, 150 mg/m³ normal dry gas.

For power plants, general sulfur dioxide emission limits are as follows:

- New coal-fired facilities 50 mg S/MJ
- Facilities emitting < 400 tS/a 100 mg S/MJ
- Facilities emitting > 400 tS/a 50 mg S/MJ.

Typical NO_x limits for different power plants are:

- Oil fired 150 mg NO_2/MJ
- Solid fuel fired 100 mg NO_2/MJ
- Gas fired 50 mg NO_2/MJ.

Waste Disposal

No general regulations apply for waste disposal. However, individual permits granted by the Licensing Board provide separate rules for waste disposal, particularly with respect to landfills. These may include, for example, regulations as to which kinds of waste may be used as landfill and which must be treated as hazardous waste and collected separately for delivery to special plants.

Economic Instruments

In 1988, Parliament decided to increase the economic instruments for implementing environmental policy. The carbon dioxide tax, the sulfur tax and the nitrogen oxide tax were introduced in 1991. The main environmental taxes and charges in 1996 are shown in Table 5[2]:

Table 5. Main environmental taxes and charges in Sweden, 1996[1].

Tax	Tax rate	Forecasted revenue 1996 (billion SEK)
Energy tax on fossil fuels	Petrol: 3.30 SEK/liter Oil for heating: 0.59 SEK/liter Industry is exempted	25.4
Carbon dioxide tax	Industry: 0.09 SEK/kg Other: 0.37 SEK/kg	14.2
Consumer tax on electricity	Industry is exempted Other: 0.097 SEK/kWh	6.8
Producer tax on electricity	Hydro: 0.04 SEK/kWh Nuclear: 0.012 SEK/kWh	1.4

Appendix 1

- Air pollution abatement

    In the beginning of 1995, the new Act on Excise Duties on Energy came into force. The new act replaced all former acts on excise duties in the field of energy taxation. In the new act, Sweden aims to follow relevant EC directives which take into consideration both the carbon dioxide content as well as the energy content in all kinds of fossil fuels.

    A sulfur tax at a rate of SEK 30 per kilogram of sulfur content is imposed on peat, coal, petroleum coke, and other solid or gaseous products. The tax was introduced with the aim of reducing emissions by 80%, using 1980 as the base year.

    A nitrogen oxide charge came into force in 1992. Only larger plants (more than 40 GWh per year) are liable to the tax. The charge is SEK 40 per kilogram of emitted nitrogen oxide. The revenue is redistributed to those plants that have reduced emissions.

- Waste disposal

    A Government Commission presents proposal for a tax on waste in autumn 1996 . The Commission will also clarify the effects of such a tax, including an assessment of its suitability as a cost-effective instrument for improved waste management.

- Use of natural resources

    A tax on natural gravel has been in force since July 1, 1996. The tax rate is SEK 5 per ton of gravel. The tax will act as an incentive to use alternative materials, especially crushed rock.

- Environmentally motivated subsidies

    The Government has a right to apply exemptions or reductions in the rates of excise duty on fuels used in pilot projects for the technological development of more environmentally-friendly products and in particular in relation to fuels from renewable sources.

Executive Authorities

The Ministry of Environment and Energy has responsibility for environmental policy. The National Environmental Protection Board is the central administrative authority in the environmental sector having comprehensive functions in connection with air and water pollution as well as noise abatement. Among other duties, the Board assists the Licensing Board for Environment Protection in surveys and field work in matters related to environment protection.

The Licensing Board for Environment Protection and the County Administration or the Municipal Authority for Environment and Health are responsible for licensing according to the Environment Protection Ordinance.

1.5 Norway

General Environmental Legislation

The main environmental law is the Pollution Control Act of 1981, which was amended in 1986. The act refers to polluting activities such as:

- Emission of solids, wastewater or gases to air, water or land
- Noise
- Light or other radiation
- Influence on the ambient temperature.

According to the Act, the pollution control authorities dictate the conditions of pollutant emission. If the polluting activity is considered as a cause of pollution, a permit must be applied for from the pollution control authorities and the permit can be granted under certain conditions. Thus the purpose of the Act is to protect the environment from pollution and reduce existing pollution.

Long-term programs for the reduction of pulp and paper industry emissions have been prepared by both the authorities and industry itself. The pulp and paper industry, for example, proposed reductions of COD and AOX loads, of 50% and 70% respectively, by 1995 with 1988 as the base level. Also, a 50% reduction in SO_2 emissions by 1993 was planned, using 1980 as the base reference.

Water Pollution Control

No general limits apply. The restrictions for each individual mill are determined on the basis of the condition of the receiving water course and the available technology. The regulated water pollutants are suspended solids, COD_{Cr}, phosphorus, and AOX.

Air Pollution Control

Air emissions from the pulp and paper industry are restricted case by case, as no general regulations have been issued. Restrictions follow the general Norwegian program for reduction of atmospheric emissions as follows:

Sulfur dioxide:	By 1993, total emissions shall be reduced by 50% based on the 1980 level.
Carbon dioxide:	Total emissions shall be stabilized at the 1989 level until 2000.
Nitrogen oxides:	By 1994, total emissions shall be reduced by 10% based on the 1987 level and by 1998, by 30% based on the 1986 level.

In the case of particulate emissions, levels of 50 mg/Nm^3 in recovery boiler flue gases and 100–200 mg/Nm^3 in bark boiler flue gases are typically allowed.

Appendix 1

Waste Disposal

Individual regulations are set for each mill. The current trend is toward incinerating organic waste, such as bark and wood residue, and collecting and special treatment of hazardous waste.

Economic Instruments

The prevailing economic instruments are differential product charges. Contrary to the other Nordic countries, Norway has an investment charge. Since 1989, the investment charge has been used as an instrument of environmental policy, as certain types of installations and equipment have been exempt from the charge[2]. This includes municipal sewage and waste treatment plants, and reception and treatment plants for hazardous waste.

- Air pollution abatement

    Charges on gasoline and car diesel are applied at present. The charge consists of a basic excise and an additional CO_2 charge.

    Charges on fuel oils, natural gas, and electricity were introduced in 1970. At present, the charge only covers sulfur and CO_2 charges, as the basic excise charge was removed in 1993.

    In 1991, a CO_2 charge on mineral oils and on natural gas and petroleum burned during production on the continental shelf, was introduced. A CO_2 charge on coal and coke used for energy purposes has been in effect since 1992.

    In addition, since the beginning of 1993, electricity has been subject to both a production and a consumption charge.

    The above charges, valid in 1996, are shown in Table 6:

Table 6. Charges of selected fuels in Norway in 1996[1].
(Note. The figures do not include VAT of 23 percent)

Fuel/Duties	Unit charge
Fuel oils	NOK/liter
- CO_2 charge	0.425
- Sulfur charge[*]	0.07
Total: light oil	0.495
Total: heavy oil	0.705–1.055
Coal/Coke	NOK/kg
- CO_2 charge	0.425
Natural gas	NOK/standard m3
- CO_2 charge	0.85
Electricity	NOK/kWh
- Electricity charge	0.053
- Production charge	0.0155

[*] *The sulfur charge is differentiated on the basis of sulfur content in the fuel oil. The rate is currently NOK 0.07 per liter for each commenced 0.25 percent sulfur content. Fuel oils with less than 0.05 percent sulfur are not liable to the sulfur charge.*

All manufacturing industry is exempt from the electricity charge. Sulfur charges may be refunded on the basis of the amount of sulfur absorbed during combustion.

- Water pollution control

    A user charge on sewage and water is applied varying from NOK 200–1000 per year. There are also charges on artificial fertilizers.

- Waste disposal

    The local authorities charge their customers for management of sewage. The user charge may not be set higher than the local authorities' costs related to these activities (range between NOK 200 and NOK 1000 per year).

- Environmentally motivated subsidies

    Subsidies for research and development are granted by the environmental authority. In 1996, NOK 20 million was allocated to a program called "Cleaner Production." In this program, environmental projects at pilot and demonstration stages are funded in order to introduce existing environment-friendly knowledge and techniques to practical projects or business.

Executive Authorities

The Ministry of the Environment has the responsibility of pollution control for the state. The Ministry also sets and coordinates the long-term objectives of pollution control.

The State Pollution Control Authority (Statens Forurensningstilsyn, SFT) is responsible for setting discharge limits for each mill. SFT also grants the discharge permits and carries out supervision to ensure that emissions are kept within the limits. SFT also administers pollutant monitoring programs and draws up new regulations and development plans.

1.6 Germany

General Environmental Legislation

In Germany, environmental legislation has been developed both on a federal government and on a state level. The key major framework laws are as follows:

- Federal Emission Control Act (BImSchG) and amendments (1974, 1985)
- 4th Ordinance of Federal Emission Control Act and amendments (1975, 1985)
- Technical Instructions on Air Quality (TA Luft, 1983, 1986)
- Federal Water Act (Wasserhaushaltsgesetz, WHG) and amendments (1986)
- Waste Management Act (Abfallgesetz, 1986)
- Wastewater Charges Act (AbwAG) and amendments (1987).

Appendix 1

TA Luft is the most important administrative regulation implementing the BIm-SchG. TA Luft stipulates detailed technical demands and emission limits for existing and new industrial plants.

The Federal Water Act, WHG, forms the framework for regulations on effluent discharges.

According to the Waste Management Act, technical directives and regulations were promulgated at the beginning of 1990s (TA Abfall).

Regulations on Effluent Discharges

The current federal minimum requirements for pulp and paper industry effluent discharge standards (Mindestandforderungen) follow the separate regulation (AbwasserVwV, Anhang 19, Teil B) shown in Table 7.

Table 7. Effluent discharge limit values for pulp and paper mills in Germany.

Production basis	TSS[b] mg/l	COD$_{Cr}$ kg/metric ton	BOD$_5$ mg/l, kg/metric ton	N[c] mg/l	P$_{tot}$ mg/l	AOX kg/metric ton
Chemical pulp						
(sulfite)	-	70	-/5	10[d]	2[e]	1
Paper and board[a]						
- Class 1	50[f]	3	25/1	10[d]	2[e]	0.04[g]
- Class 2	50[f]	6	25/2	10[d]	2[e]	0.04[g]
- Class 3	50[f]	9	25/3	10[d]	2[e]	0.04[g]
- Class 4	50[f]	12	-/6	-	2[e]	0.025[g]
- Class 5	-	2	25/-	10[d]	2[e]	0.02[g]
- Class 6	-	3(5)[h]	25/-	10[d]	2[e]	0.01[g]
- Class 7	-	5	25[i]/-	10[d]	2[e]	0.012[g]

[a] Paper and board mills are divided into the following groups:
  1 Woodfree unsized
  2 Woodfree sized
  3 Woodfree, highly refined and special paper (with more than one quality change per working day as annual average)
  4 Pergament
  5 Woodfree, coated, more than 10 g coating/m^2 (intregrated)
  6 Woodcontaining (integrated with mechanical pulping), (integrated) end product predominantly not from
  7 Recycled paper, mainly based on recycled fiber

[b] Filterable solids
[c] Ammonium-, nitrate- and nitrite-N
[d] When effluent amount exceeds 500 m^3/d
[e] When effluent amount exceeds 1000 m^3/d
[f] In cases where effluent is subjected to biological teratment
[g] At specific conditions regarding the use of chlorohydrine- containing wet-strength papers the limit value is 0.12 kg/metric ton or 0.2 kg/metric ton
[h] 5 kg/metric ton when over 50% of the pulp is TMP or when a substantial part of the pulp is bleached with hydrogen peroxide
[i] If effluent amount is below 10 m^3/metric ton, the limit value is 50 mg/l and the specific limit value 0.25 kg/metric ton, respectively

Regulations on Atmospheric Emissions

The Ministry of the Environment sets restrictions on air emissions in the Technical Regulations for Air Pollution Control (Technische Anleitung zur Reinhaltung der Luft, TA-Luft 1986). Table 8 presents these general restrictions.

Table 8. Limit values for atmospheric emissions from industrial plants in Germany (1986).

Substance	Limit value mg/m^{3}n	Resp. absolute emission
Particulates	50	> 0.5 kg/h
	150	< 0.5 kg/h
Hydrogen sulfide (H$_2$S)	5	> 50 g/h
Sulfur oxides as SO$_2$	500	> 5 kg/h
Nitrogen oxides as NO$_2$	500	> 5 kg/h
Chlorine (Cl$_2$)	5	> 50 g/h

In addition, emission limits are set for a number of metal compounds (e.g., Cd, Hg, Ni, Pb, Cu, Cr, Mn) and for certain other substances, e.g., various volatile organic compounds.

Depending on the size and type of the power plants and the fuel used, special additional regulations are also applied.

Regulations on Waste Disposal

Disposal and handling of various wastes is at present tightly controlled in Germany. Important recent regulations are the following:

- 2nd General Administrative Regulation implementing the Waste Management Act (Zweite allgemeine Verwaltungsvorschrift zum Abfallgesetz, TA-Abfall, 1991)

- Regulation on incineration plants for wastes and similar combustible substances (verordnung uber Verbrennungsanlagen fur Abfälle und ähnliche brennbare Stoffe - 17. BlmSchV, 1991)

- Wastewater sludge regulation (Klärschlammverordnung–AbfKlärV, 1992).

Special attention is paid in TA-Abfall to the characteristics of waste to be disposed of at landfills; for example, there are regulations specifying the volatile matter content, conductivity, TOC, Cd, Cr, Cu, and water-soluble substances of the waste.

Table 9 shows the current regulations for waste incineration plants.

Appendix 1

Table 9. Limits for atmospheric emissions from waste incineration plants in Germany (1991).

Compound	Average daily limit, mg/m³n	Average 1/2 hour limit, mg/m³n
Particulates	10	30
Organic compounds, as total carbon	10	20
Inorganic chlorine gaseous compounds, as HCl	10	60
Inorganic gaseous fluorine compounds, as HF	1	4
Sulfur oxides as SO_2	50	200
Nitrogen oxides as NO_2	200	400

Note. Additional limit values exist for certain metals (e.g. Cd, Hg, Pb, Cr, Mn, Ni) and for certain dioxines and furanes.

Note. Additional limit values exist for certain metals (e.g. Cd, Hg, Pb, Cr, Mn, Ni) and for certain dioxines and furanes.

The Wastewater sludge regulation sets a number of restrictions on the landfill disposal of waste sludge (e.g., permitted types, characteristics, and sites).

Economic Instruments

According to Wastewater Charges Act amendment (AbwAG, 1991), industrial plants are liable to pay discharge fees if certain limit values are exceeded. Table 10 shows the damage units (Schade einheit) for determination of discharge fees.

Table 10. Damage units for determination of discharge fees, applicable to pulp and paper mills in Germany (1991).

Pollutant	Damage unit	Limit value (as concentration or annual load)
COD_{Cr}	50 kg	20 mg/l, 250 kg/year
AOX	2 kg	100 ug/l, 10 kg/year
Phosphorus, as P	3 kg	0.1 mg/l, 15 kg/year
Nitrogen, as N	25 kg	5 mg/l, 125 kg/year
Acute toxicity to fish		
- effluent amount	3000 m³	$G_Fa = 2$
Metals and their compounds		
- Hg	20 g	1 ug/l, 100 g/year
- Cd	100 g	5 ug/l, 500 g/year
- Cr	500 g	50 ug/l, 2.5 kg/year
- Ni	500 g	50 ug/l, 2.5 kg/year
- Pb	500 g	50 ug/l, 2.5 kg/year
- Cu	1000 g	100 ug/l, 5 kg/year

$^a G_F$ is the dilution factor by which the waste water turns to nontoxic to fish, according to a specified test method.

The present fee (since Jan. 1, 1997) per damage unit is DEM 80. From 1999, the fee will be increased to DEM 90.

Executive Authorities

The Federal Republic is the supreme authority for legislation and its representative is the Federal Ministry of the Environment (Bundesministerium fur Umwelt, Naturschutz und Reaktorsicherheit, BMU).

The Federal States (Bundesländer) implement the Acts via the BMU, but they can also promulgate stricter emission limits than the federal standards.

The environmental permits are granted by state administrative expert organs.

1.7 France

General Environmental Legislation

Major environmental laws in France are as follows:

- The Act on Air Pollution and Odor Abatement (1961)
- The Act on Installations Registered for Environmental Protection (1976)
- Instructions Regarding the Paper Industry (1976)
- Technical Instructions for the Paper and Board Industry (1989).

On Jan. 6, 1994, the Ministry of the Environment issued new regulations for the pulp and paper industry, which are currently in force. These regulations are presented below.

Regulations on Effluent Discharges

The 1994 decree (Arrêté Ministériel) specifies some general requirements for the quality of effluents to be released into watercourses. Main requirements are as follows:

- pH to be maintained between 5.5 and 8.5
- Temperature to be below 30°C (35°C if anaerobic wastewater treatment is used)
- Color to be less than 100 mg Pt/l
- Phenolic type substances 0.3 to be less than mg/l or 3 g/d
- Phenols to be less than 0.1 mg/l or 1 g/d
- AOX to be less than 5 mg/l or 30 g/d
- Total hydrocarbons to be less than 10 mg/l or 100 g/d
- Additional specific requirements for toxic and bioaccumulative substances are listed in Annex 4 of the decree (several organic and inorganic compounds).

For pulp mills, Table 11 shows the current regulations for new plants and Table 12 shows those for existing plants.

Appendix 1

Table 11. Effluent discharge limit values for new pulp mills in France. The values are expressed as kg/a.d. metric ton, monthly average max. values.

Mill type	TSS	BOD$_5$	COD$_{Cr}$
MECHANICAL			
- unbleached	0.7	0.7	1.5
- bleached		0.7	0.7	3.0
TMP			
- unbleached	0.7	0.7	4.5
- bleached		0.7	0.7	6.0
CTMP			
- unbleached	0.7	3.0	12.0
- bleached		0.7	4.0	16.0
KRAFT (hardwood)			
- unbleached	5.0	1.5	15.0
- bleached		5.0	2.0	25.0
KRAFT (softwood)			
- unbleached	5.0	2.0	20.0
- bleached		5.0	3.0	50.0
BISULFITE		5.0	5.0	35.0
RECYCLED PAPER (deinking)	0.7	0.7	4.0

Table 12. Effluent discharge limit values for existing pulp mills in France. The values are expressed as kg/a.d. metric ton, monthly average max. values.

Mill type	TSS	BOD$_5$	COD$_{Cr}$
MECHANICAL			
- unbleached	0.9	0.9	2.0
- bleached		0.9	0.9	3.9
TMP			
- unbleached	0.9	0.9	5.9
- bleached		0.9	0.9	7.8
CTMP			
- unbleached	0.9	3.9	15.6
- bleached		0.9	5.2	20.8
KRAFT (hardwood)			
- unbleached	6.5	2.0	19.5
- bleached		6.5	2.6	32.5
KRAFT (softwood)			
- unbleached	6.5	2.6	26.0
- bleached		6.5	3.9	65.0
BISULFITE		6.5	6.5	45.5
RECYCLED PAPER (deinking)	0.9	0.9	5.2
AOX restriction for bleached pulp mills is max. 1 kg/a.d. metric ton.			

For paper mills, Table 13 shows the current restrictions for new plants and Table 14 shows those for existing plants.

Table 13. Effluent discharge limit values for new paper mills in France. The values are expressed as kg/a.d. metric ton, monthly average max. values.

End product	TSS	BOD$_5$	COD$_{Cr}$
Paper with more than 90% virgin fiber, without fillers	0.7	0.7	2.5
Paper with more than 90% virgin fiber, with fillers or coating	0.7	0.7	3.0
Paper with more than 90% virgin fiber with fillers and coating	0.7	0.7	3.0
Paper with more than 90% recovered paper, without fillers	0.7	0.7	3.0
Paper with more than 90% recovered paper, with fillers or coating	0.7	0.7	4.0
Paper with more than 90% recovered paper, with fillers and coating	0.7	0.7	4.0

Table 14. Effluent discharge limit values for existing paper mills in France. The values are expressed as kg/a.d. metric tons, monthly average max. values.

End product/capacity	TSS	BOD$_5$	COD$_{Cr}$
Capacity inferior to 60 a.d. metric ton/d	2.0	4.0	8.0
Paper with more than 90% virgin fiber, without fillers	1.5	1.0	4.0
Paper with more than 90% virgin fiber, with fillers or coating	1.5	1.5	6.0
Paper with more than 90% virgin fiber with fillers and coating	1.5	2.0	8.0
Paper with more than 90% recovered paper, without fillers	1.5	1.5	6.0
Paper with more than 90% recovered paper, with fillers or coating	1.5	2.0	8.0
Paper with more than 90% recovered paper, with fillers and coating	1.5	2.0	8.0
Fluting	1.9	1.9	8.0

In each case, the daily maximum value may be twice as high as the monthly average maximum value in the above-mentioned tables.

Appendix 1

Regulations on Atmospheric Emissions

According to the 1994 decree, the current atmospheric restrictions are:
   Flue gas unit volumes referred to as m³n represent volumes in normal conditions: i.e., temperature 273 °K, pressure 101.3 kPa, dry gas, with 6% vol. oxygen content.

- Total particulates:	80 mg/m³n for recovery boilers 100 mg/m³n for lime kilns 50 mg/m³n for other installations
- Carbon monoxide (CO):	To be restricted locally
- Sulfur oxides (as SO_2):	If the mass flow is superior to 25 kg/h, the general limit is 300 mg/m³n. For bisulfite pulping, the limit value is 500 mg/m³n, respectively.
- Nitrogen oxides (as NO_2):	If the mass flow is superior to 25 kg/h, the limit is 500 mg/m³n.
- Chlorine (as HCl):	If the mass flow is superior to 1 kg/h, the limit is 50 mg/m³n.
Total organic compounds, except methane	If the mass flow is superior to 2 kg/h, the limit is methane150 mg/m³n. If the effluents are incinerated, the limit is 50 mg/m³n, expressed as total C.
Organic compounds, as listed in Annex III (e.g., aniline, biphenyls, chloroform, cresol, dimethylamine, nitrotoluene, phenol, pyridine):	If the mass flow exceeds 0.1kg/h, the limit is 20 mg/ m³n.

No numerical restrictions for **odor control** exist, only general recommendations to decrease odor release (in effluent treatment, for example) are presented.

Regulations on Waste Disposal

Waste disposal arrangements are defined in the Law of July 19, 1976, including principles concerning recycling, handling, and temporary storage of waste. In the event that the waste cannot be recycled or reused, it must be removed to authorized installations.

   On the basis of the current waste regulations of 1985 and 1990, a separate waste evaluation must be carried out for each new industrial installation. The study must include a description of the waste (origin, quantity, quality, storing, end use), waste management, and technical and economical evaluation of waste handling.

The authorities aim at rapid decreasing of wastes to be disposed of as landfill toward the end of the 1990s. After July 1, 2002, only "ultimate" waste will be allowed to be disposed of as landfill (Law of July 15, 1975).

Executive Authorities

Each pulp or paper mill is subject to a local operating license (Arrêté Préfectoral). The licenses must comply with the new national legislation within one year for new mills and within four years for existing mills.

1.8 United Kingdom

General Environmental Legislation

The main laws currently in force are as follows:

- The Environmental Protection Act (EPA) (1990)
- The Environmental Protection (Prescribed Processes and Substances) Regulations (1991) and Amendments (1994, 1995).

Part I of the Environmental Protection Act (EPA) 1990 introduced a new system of Integrated Pollution Control (IPC) to protect against the release of certain prescribed substances into the environment (water, air, land). Part I of the EPA also requires the operators of prescribed processes to apply for authorization. From April 1991, all new and substantially altered major installations have been subject to IPC.

According to the Environmental Protection Regulations the main classes of prescribed processes are as follows:

- Fuel and power, including petroleum
- Waste disposal (incinerators, chemical recovery, waste derived fuel but not landfill sites)
- Minerals (asbestos, cement, fiber, glass, ceramics)
- Chemicals, including petrochemicals, fertilizers, and pharmaceuticals
- Metals
- "Other," including paper, timber, tar, and coatings.

The above-mentioned industries must apply to Her Majesty's Inspectorate of Pollution (HMIP) for authorisation. HMIP was established in 1987 and since April 1, 1996, has been part of the Environment Agency (Department of the Environment).

The Regulations also give a list of prescribed substances, the release of which into air, water, or land must be prevented or minimized. Plants subject to IPC are required to apply the Best Available Technique Not Entailing Excessive Cost (BATNEEC) to control emissions.

Appendix 1

Regulations on Effluent Discharges

The current laws and regulations concerning control of water pollution are:

- The Water Resources Act 1991
- The Water Industry Act 1991
- The EPA 1990.

Effluent discharges to water courses are controlled by the Water Resources Act. The Act establishes Water Quality Objectives (WQOs). The Secretary of State must establish WQOs for all relevant territorial waters, coastal waters, and inland freshwaters (surface and ground waters). It is the joint responsibility of the Secretary of State and the National Rivers Authority (NRA) to see that these objectives are achieved at all times.

The Water Resources Act also requires that discharges of polluting substances are in accordance with the terms of a discharge consent, issued by the NRA.

According to the IPC principle, limit values of certain prescribed substances for release into water are determined in the Regulations (Amendments 1994, 1995). The limit value is expressed as the "amount in excess of background quantity released in any 12-month period" for certain heavy metals and organic compounds.

Emission limits for point sources, e.g., industrial plants, are determined to maintain the quality objectives of the receiving water, taking into account all other discharges to the same water course as well. Typically, regulations are set for wastewater flow, TSS, BOD, COD, and pH. These requirements vary to some extent depending on the type of the receiving water (e.g., river, inland lake, estuary).

Discharges of wastewater to public sewers are controlled by the Water Industry Act and the Trade Effluents (Prescribed Processes and Substances) Regulations 1989. A consent is also required, when wastewater is released into a common sewer.

Regulations on Atmospheric Emissions

Basic main laws are:

- Clean Air Act 1993
- Environmental Protection (Prescribed Processes and Substances) Regulations 1991
- The EPA 1990.

In air pollution control, the IPC principle is being introduced in stages, and will eventually apply to all processes in England and Wales. The Regulations 1991 prescribe substances whose release into air should be controlled, as follows:

- Oxides of sulfur and other sulfur compounds
- Oxides of nitrogen and other nitrogen compounds

- Oxides of carbon
- Organic compounds and products of partial oxidation
- Metals, metalloids, and their compounds
- Asbestos (suspended particulate matter and fibers), glass fibers, and mineral fibers
- Halogens and their compounds
- Phosphorus and its compounds
- Particulate matter.

The Regulations 1991 provides two lists – one of processes under HMIP integrated pollution control (Part A), the other subject to local authority air pollution control (Part B). The Secretary of State has issued guidance, e.g., on the way in which local authorities (District Councils) are to process applications and establish public registers.

At present, only chemical pulp mills (production more than 25,000 metric tons/year) are classified in Part A and under HMIP. All nonintegrated paper mills belong to Part B and are under the control of District Councils.

According to the Clean Air Act, the local authorities can designate a certain area as a Smoke Control Area, and apply stricter emission limits.

Regulations on Waste Disposal

The current key laws and regulations are:

- Collection and Disposal of Waste Regulations 1988
- The EPA 1990
- Controlled Waste Regulations 1992.

According to IPC, the release of the following prescribed substances onto land must be eliminated or controlled:

- Organic solvents
- Azides
- Halogens and their covalent compounds
- Metal carbonyls
- Organo-metallic compounds
- Oxidizing agents
- Polychlorinated dibenzofuran and any congener thereof
- Polychlorinated dibenzo-p-dioxin and any congener thereof

Appendix 1

- Polyhalogenated biphenyls, terphenyls and naphthalens
- Phosphorus
- Pesticides
- Alkali metals and their oxides and alkaline earth metals and their oxides.

The EPA 1990 introduced Waste Regulatory Agencies (WRAs) to implement IPC procedures. At local level, Local Authority Waste Disposal Companies (LAWDCs) were established to operate waste disposal, treatment, and storage facilities.

Disposal of "controlled" waste is allowed only at sites holding a disposal license, according to the EPA 1990. "Controlled" waste means household, industrial, or commercial waste, excluding agricultural, mining, and quarry waste.

A duty of care is placed on those dealing with all other waste, except household waste disposed of separately. Under the duty of care, waste may be transferred to authorized persons only.

1.9 Italy

General Environmental Legislation

The following framework environmental laws have been issued:

- Law for Water Pollution Control (319/1976) 1976
- Law for Air Pollution Control (615/1966) and Supplements 1971
- Law for Solid Waste Disposal (915/1982) and Supplements 1987, 1988.

In addition, most of the EU's environmental directives and regulations have been adopted in Italy or are under implementation.

Regulations on Effluent Discharges

According to the 1976 Law, wastewater standards are divided into three categories:

- Tabella A: Industrial wastewaters when discharged directly into a water course
- Tabella B: Industrial wastewaters when discharged to a municipal sewage treatment plant
- Tabella C: Municipal sewage (over 50 PE).

Table 15 shows limit values according to Tabella A.

Table 15. Industrial effluent limits in Italy according to Tabella A (direct release to water course).

Object	Limit value	Remarks
pH	5.5–9.5	After dispersion, 50 m from the discharge point 6.5–8.5
Total suspended solids (TSS)	80 mg/l	Filter openings 0.45 um
Settleable solids	0.5 ml/l	Imhoff cone, 2 hours
BOD_5	40 mg/l	For certain industrial effluents, the limit may be equal to 70% of the total BOD_5 release
COD_{Cr}	160 mg/l	
Temperature increment	< 3 °C	In rivers, after the dispersion zone
Colour	non-visible	When dilution is 1:20
Total phosphorus	10 mg/l	In certain lakes and dams 0.5 mg/l

Regulations on Atmospheric Emissions

The general limits for air emissions are:

- Sulfur oxides (as SO_2)   500 mg/m³n
- Nitrogen oxides (as NO_2)   50 mg/m³n (as guideline)
- Total reduced sulfur, TRS   5 mg/m³n

At present, all EU regulations on air pollution control are in force or under implementation in Italy.

Regulations on Waste Disposal

There are regulations for toxic waste collection and recycling of solid waste such as paper and glass.

1.10 Spain

General Environmental Legislation

The basic environmental laws are:

- Law of Water 1985
- Law of Atmospheric Pollution 1972
- Laws of Solid Wastes 1975 and 1986.

Spanish environmental legislation and authorities follow closely the relevant EU regulations. The major part of the EU environmental decrees has been adopted in Spain or is under implementation.

Appendix 1

Regulations on Effluent Discharges

General limit values for industrial effluent discharges are given in the 1985 Law. These limit values also form a basis for a discharge fee (tax) system. The annual fee (F) is determined by the following equation:

$$F = C \times P$$

where C is annual contaminating units
       P contaminating unit price

$$C = K \times Q$$

where C is annual contaminating units
       K quality factor, dependent on the pollutants in the effluent
       Q annual effluent flow in m^3 (Note: "clean" waters, e.g., from cooling units, can be excluded).

The discharge tax system is applied to mills which release their effluents into rivers or lakes.

Table 16 shows limit values divided into three categories.

Table 16. Effluent discharge limit values for calculation of discharge fees in Spain.

Subject	Category 1	Category 2	Category 3
pH[a]	5.5–9.5	5.5–9.5	5.5–9.5
Suspended solids, mg/l[b]	300	150	80
Settleable solids, ml/l[c]	2	1	0.5
Coarse solids	none	none	none
BOD$_5$, mg/l[d]	300	60	40
COD$_{Cr}$, mg/l	500	200	160
Temperature increment, °C[e]	3	3	3
Colour, nonvisible[f]	1:40	1:30	1:20
Total phosphorus, mg/l[g]	20	20	10

[a]After dispersion, 50 from the discharge point resp. level 6.5–8.5
[b]Filter openings 0.45 μm
[c]Imhoff cone, 2 h settling
[d]For certain industrial effluents, the limit may be equal to 70% of the total BOD$_5$ release.
[e]In rivers, after dispersion zone. In lakes, max. effluent temperature is 20 °C.
[f]Color to be determined through 10 cm of diluted effluent in each category.
[g]In certain lakes and dams, the limit value is 0.5 mg/l.

The authorities determine the quality factors for each category. The most polluting substance as specified in Table 16 determines the category and quality factor to be used.

In addition to the pollutants shown in Table 16, limit values for certain metals and their compounds and organic compounds can also be applied.

The general minimum requirements have been issued primarily for use in the discharge tax system. Additional regulations can be set up, based on the type and use of the receiving water, for example.

Regulations on Atmospheric Emissions

Ambient air quality standards based on EU regulations were adopted during the 1970s and 1980s.

Air emission limits applicable to the pulp and paper industry are as follows:

- Particulates: < 150 mg/m^{3}n for recovery boilers and power boilers
- Sulfur dioxide: < 2400 mg/m^{3}n for industrial coal fired power boilers
  < 5 kg/t for sulfite pulping
- Hydrogen sulfide: < 7.5 mg/m^{3}n for recovery boilers
  < 10 mg/m^{3}n general limit
- Nitrogen oxides (NO$_2$): < 300 ppm
- Carbon monoxide (CO): < 500 ppm
- Hydrogen chloride (HCl): < 460 mg/m^{3}n

Regulations on Waste Disposal

No specific regulations are applied to the pulp and paper industry. Waste directives issued by the EU are to be followed for classification of waste into toxic and nontoxic groups, etc.

Environmental Impact Assessment (EIA)

In 1988, following the Royal Decree 1131, EIA procedure became obligatory for new pulp and paper mill investments, for example.

Economic Instruments

The discharge fee system is applied for water pollutants. The monetary value of a contaminating unit is determined for a four-year period.

Executive Authorities

The General Secretary of the Environment in the Ministry of Public Works and Urban Development (MOPU) is the responsible government agency for environmental policy

Appendix 1

and coordination. Major administrative tasks have been assigned to the 17 Provincial Administrations, (Comunidades Autónomas). Ten river basin confederations control in water supply and release of wastewater.

1.11 Portugal

General Environmental Legislation

In Portugal, environmental legislation has been developed mainly during the 1980s and 1990s on the basis of EU legislation. In 1988, the pulp and paper industry signed an agreement with the Ministry for Industry (DGI) and with the General Directorate for Environmental Quality (DGQA) on incorporating the EU regulations into Portuguese legislation.

In 1992, the Ministry of the Environment proposed an Environmental Pact with industry, including the pulp and paper industry, to achieve a 50%–60% reduction in the total pollution caused by industrial facilities. The program emphasizes the importance of the use of low-pollution technology in rebuilds and new mills to reach acceptable pollution loads on water, air, and soil.

Regulations on Effluent Discharges

Table 17 lists the current limit values (1996) for kraft mills and paper mills.

Table 17. Effluent discharge limit values in Portugal 1996.

Mill	TSS	BOD_5	COD_{Cr}
Unbleached kraft	1.5	3	35
Kraft liner (integrated)	2	4	30
Nonintegrated paper	60 mg/l	40 mg/l	150 mg/l

Regulations on Atmospheric Emissions

Table 18 shows emission limits for kraft mills.

Table 18. Air emission limits for kraft mills in Portugal.

Source	Air emission, mg/m³n[a]		
	Particulates	H_2S	SO_2
Recovery boiler	150	10	500
Lime kiln	200	50	–
Power boiler			
- coal fired	150	–	1700
- biofuel	300	–	–

[a] dry gases, 8% O_2

Regulations on Waste Disposal

Current waste disposal regulations mostly follow EU directives. Special treatment is required for oil waste and waste containing PCB, etc.

Executive Authorities

The Ministry of Industry and the Ministry of the Environment provide for environmental policy and implementation of environmental legislation. The General Directorate for Environmental Quality (DGQA) acts as the control and follow-up agency in environmental discharge permit procedures.

1.12 The United States

General Environmental Legislation

In the United States, both federal and state levels develop environmental legislation.
At the federal level, the major framework laws are:

- The National Environmental Policy Act (NEPA) 1970
- The Clean Air Act (CAA) 1970 and amendments
- The Clean Water Act (CWA) 1972 and amendments
- The Toxic Substances and Control Act (TSCA) 1976 and amendments
- The Resource Conservation and Recovery Act (RCRA) 1976
- The Comprehensive Environmental Response, Compensation and Liability Act (CERCLA) 1980
- The Superfund Amendments and Reauthorization Act (SARA) 1986.

The NEPA established the United States Environmental Protection Agency (EPA) for the development and enforcement of environmental regulations.

The Clean Air Act (CAA) required the EPA to establish ambient air quality standards which would be the basis for determining the appropriate regulations on industrial pollution sources. The primary responsibility for enforcing these regulations was intended to be taken by the individual states, under EPA supervision. Standards were also to be set for new pollutant sources ("New Source Performance Standards" or NSPS), based on the ability of current technology to minimize pollution.

The CWA charged the EPA with setting technology-based effluent regulations requiring industry to meet pollution control requirements equivalent to the "Best Practical Technology" (BPT) and the "Best Available Technology Economically Achievable" (BAT) by 1983. It was the task of the EPA to interpret these acronyms for each branch of industry and promulgate specific regulations. This Act also authorized the EPA to oversee state standards for ambient water quality.

Subsequent amendments to the CAA and the CWA have further incorporated elements of both abatement philosophies (ambient quality-based and technology-based standards). Other pollution control strategies are also incorporated, such as listing the

Appendix 1

toxic chemicals which must be regulated. The CAA amendments 1990 specify 189 air toxicants for which new regulations must be developed.

Several laws have been passed which deal with solid waste. The TSCA was adopted in 1976 to regulate the production, transportation and disposal of toxic chemicals. The RCRA is concerned with the shipment and disposal of solid and hazardous waste.

The CERCLA was passed in 1980 in order to provide a mechanism for the cleanup of accidental hazardous discharges (solid, liquid, or airborne). It established a fund to be raised through a tax on chemical production which would be used for any necessary cleanups. A related Act, the SARA, passed in 1986 to strengthen the regulatory program and increase public access to information on any releases.

Individual states are free to pass their own environmental legislation and develop their own standards, which may be more (but not less) stringent than the federal standards. State legislation often mimics federal legislation by establishing local versions of the EPA. These local agencies take over the regulatory duties as long as they are judged to be at least as effective as an EPA-administered program.

In all legislation, an important role is reserved for public participation in the process and access to information on emissions and compliance records. Permits and violations are all on public record.

The Cluster Rules (1998)

The EPA has issued a revision of current environmental regulations applicable to pulp and paper operations. As both air and water regulations for pulp and paper mills are being revised concurrently in an integrated approach, the revision is called the "Cluster Rule."[6]

The first proposal for the Cluster Rule was issued on Dec. 17, 1993, and covered new effluent and air emission limitations. After a period reserved for commenting by industry and various organizations, the EPA issued revised proposals and supplemental documents during 1996. New effluent limitations for bleached, paper-grade chemical pulps (kraft and sulfite) were promulgated in April 1998. Limitations for a further ten production categories will be issued later. Air regulations for all chemical pulp mills were finalized concurrently. Additional air regulations applicable to chemical recovery processes were proposed in April 1998 and are scheduled for promulgation in April 1999.

Regulations on Effluent Discharges

The EPA issues technology-based guidelines at a federal level for effluent limit values on effluent discharges. These are mainly adopted as such in state legislation and incorporated into mill permits.

Various technological concepts are determined as a basis for the requirements of the effluent discharge regulations. These are defined as follows:

- BPT (Best Practicable Control Technology) effluent guidelines apply to discharges of conventional pollutants (BOD5, TSS, pH, oil, and grease).

- BCT (Best Conventional Pollutant Control Technology) is at present equal to BPT.
- BAT (Best Available Technology) standards define direct discharge limits for toxic and non-conventional pollutants. Prior to the Cluster Rule, the BAT concept had limited significance as it only concerned mills which use chlorophenol-based biocides. The effluent guidelines promulgated as part of the Cluster Rule establish limits for AOX, chloroform, dioxin, furan and 12 chlorinated phenolics.
- NSPS (New Source Performance Standards) are applied to new or rebuilt mills discharging their effluents directly to watercourses. NSPS regulations include conventional, nonconventional, and toxic pollutants.
- PSES and PSNS (Pretreatment standards for existing and new sources) are specific standards for discharges from existing and new mills to publicly-owned sewage treatment works.

In addition, the Cluster Rule includes the following definition for all mills:

- BMP (Best Management Practices) for spent pulping liquor management and spill prevention and control in chemical pulp mills.

The current BPT effluent discharge restrictions were adopted in 1982 and were determined for 24 mill types. Table 19 gives examples of effluent restrictions for major production groups.

Table 19. Effective EPA effluent restrictions for six production groups, expressed as annual averages, kg/a.d. metric tons, in the United States.

Mill type/Group	BPT BOD_5	TSS	NSPS BOD_5	TSS
A Unbleached kraft and paper	2.8	6.0	1.8	3.0
B Bleached market kraft pulp	8.05	16.4	5.5	9.5
H Bleached kraft pulp and fine paper	5.5	11.9	3.1	4.8
J Bleached sulfite pulp and paper	16.5	23.5	2.36	3.03
L TMP and paper	5.55	8.35	2.5	4.6
N Groundwood pulp and newsprint	3.9	6.85	2.5	3.8

Permitted pH range of treated effluent is 5-9. In wet debarking, an additional BOD load of 1.2 kg/a.d. metric tons and a TSS load of app. 3.1 kg/a.d. metric tons is permitted. The NSPS restrictions also include loads from debarking plant.

State rgulations can be tighter and may also include other parameters (e.g., AOX, color, dioxins/furans, heavy metals).

In the **Cluster Rule** proposals, the production groups were reduced to 12 in number and new BAT restrictions for bleached Kraft, soda and sulfite mills were set. General technology concepts according to the **Cluster Rule** are:

- **BAT** (existing mills): 100% ClO_2 substitution
- **NSPS** (new mills): 100% ClO_2 substitution + oxygen delignification or extended cooking. Kappa value of pulp to bleaching: softwood 14, hardwood 10.

Appendix 1

Effluent restrictions including (for bleaching plant effluents) were set as follows:

SUMMARY OF EPA's
November 14, 1997, REVISED EFFLUENT LIMITATIONS GUIDELINES

I. Overview

Table 20. New Effluent Limitations Guidelines and Pretreatment Standards for Existing, non-TCF* Bleached Kraft or Soda papergrade pulp Mills.

Compound	Monthly Ave. Point/Frequency	Daily Max.	Compliance	Comments
Effluent Limitations Guidelines for Direct Discharging Mills				
TCDD	Not Specified	ML**	Bleach Plant/ Monthly	
TCDF	Not Specified	31.9 ppq	Bleach Plant/ Monthly	
12 Chlorinated Phenolics	Not Specified	<ML**	Bleach Plant/ Monthly	
Chloroform	4.14 gm/ton	6.92 gm/ton	Bleach Plant/ Weekly	EPA Requests Comment On Certification Option
AOX	0.623 kg/ton	0.951 kg/ton	Final Effluent/Daily	
COD	Reserved	Reserved	Final Effluent	Permit writers Encouraged to write BPJ limits
BOD/TSS	Same as Current Limits	Same as Current Limits		
Pretreatment Standards for Mills that Discharge to POTWs				
TCDD	Not Specified	<ML**	Bleach Plant/ Monthly	
TCDF	Not Specified	31.9 ppq	Bleach Plant/ Monthly	
12 Chlorinated Phenolics	Not Specified	<ML**	Bleach Plant/ Monthly	
Chloroform	4.14 gm/ton	6.92 gm/ton	Bleach Plant/ Weekly	EPA RequestsComments On Certification Option
AOX	1.41 kg/ton	2.64 kg/ton	Bleach Plant/Daily	
COD	Reserved	Reserved	Assumed to be the Discharge to the POTW	EPA indicates that local limits should be imposed if COD passes through or Interferes with POTW
BOD/TSS	None	None		

* TCF = Totally Chlorine Free, meaning that no chlorine-containing chemicals are used in pulp bleaching
** ML = Minimum Level for the Analytical Method, See Table 21

Table 21. Minimum Levels for Regulated Compounds.

Compound	Minimum Level	EPA Test Method	Compound	Minimum Level	EPA Test Method
2,3,7,8-TCDD	10 ppq (pg/L)	1613	4,5,6-Trichloroguaiacol	2.5 ppb (µg/L)	1653
2,3,7,8-TCDF	10 ppq (pg/L)	1613	Tetrachloroguaiacol	5.0 ppb (µg/L)	1653
2,4,6-Trichlorophenol	2.5 ppb (µg/L)	1653	3,4,6-Trichlorocatechol	5.0 ppb (µg/L)	1653
2,4,5-Trichlorophenol	2.5 ppb (µg/L)	1653	3,4,5-Trichlorocatechol	5.0 ppb (µg/L)	1653
2,3,4,6-Tetrachlorohenol	2.5 ppb (µg/L)	1653	Tetrachlorocatechol	5.0 ppb (µg/L)	1653
Pentachlorophenol	5.0 ppb (µg/L)	1653	Trichlorosyringol	2.5 ppb (µg/L)	1653
3,4,6-Trichloroguaiacol	2.5 ppb (µg/L)	1653	AOX	20 ppb (µg/L)	1650
3,4,5,-Trichloroguaiacol	2.5 ppb (µg/L)	1653			

Table 22. New Effluent Limitations Guidelines and Pretreatment Standards for Existing Sodium-, Calcium-, or Magnesium-Based Papergrade Sulfite Mills. (Except those producing specialty grades).

Compound	Monthly Ave.	Daily Max.	Compliance Point/Frequency
Effluent Limitations Guidelines for Direct Discharging Mills			
AOX	Not Specified	<ML (see Table 21)	Final Effluent/None Specified
COD	Reserved	Reserved	
BOD/TSS	Same as Current Limits	Same as Current Limits	
Pretreatment Standards for Mills that Discharge to POTW			
AOX	Not Specified	<ML (see Table 21)	Bleach Plant/None Specified
COD	Reserved	Reserved	
BOD/TSS	None	None	

Appendix 1

Table 23. New Effluent Limitations Guidelines and Pretreatment Standards for Existing Ammonium-Based or Specialty Papergrade Sulfite Pulp Mills.

Compound	Monthly Ave.	Daily Max.	Compliance Point/Frequency
Effluent Limitations Guidelines for direct discharging Mills			
TCDD	Not Specified	<ML*	Bleach Plant/Monthly
TCDF	Not Specified	<ML*	Bleach Plant/Monthly
12 Chlorinated Phenolics	Not Specified	<ML*	Bleach Plant/Monthly
Chloroform	Reserved	Reserved	
AOX	Reserved	Reserved	
COD	Reserved	Reserved	
BOD/TSS	Same as Current Limits	Same as Current Limits	
Pretreatment Standards for Mills that Discharge to POTW			
TCDD	Not Specified	<ML*	Bleach Plant/Monthly
TCDF	Not Specified	<ML*	Bleach Plant/Monthly
12 Chlorinated Phenolics	Not Specified	<ML*	Bleach Plant/Monthly
Chloroform	Reserved	Reserved	
AOX	Reserved	Reserved	
COD	Reserved	Reserved	
BOD/TSS	None	None	

*ML = Minimum Level for the Analytical Method, See Table 21

Regulations on Atmospheric Emissions

In connection with the Amendment of the Clean Air Act in 1990, the EPA has set regulations for various emissions into air. Definitions concerning air pollution control in the EPA regulations are:

- **NSPS** (New Source Performance Standards) include limit values of TRS and particulate emissions from new kraft pulp mills.

In the Cluster Rule additional definition is given:

- **MACT** (Maximum Achievable Control Technology) regulates the emissions of hazardous air pollutants (HAP) released in pulp and paper mills. **MACT** standards are at present divided into the following groups:

- **MACT I**: control of HAP from pulping, bleaching and wastewater emission points in new and existing chemical pulp mills

- **MACT II**: control of HAP from other emission sources (combustion sources) in new and existing chemical pulp mills
- **MACT III**: control of HAP emission from mechanical pulping, secondary fiber and paper machines.

The EPA issued a first proposal for MACT I standards in December 1993 and a revised proposal on March 8, 1996. Final MACT I Standards were published April 15, 1998. MACT II standards were issued as a proposal in April 1998 and expected to become final rules in April 1999. The MACT III standards were issued as proposed standards on March 8, 1996 and finalized in April 1998. All standards must be met with 3 years of the rule becoming final. One exception is that air emissions from brown stock washer systems must be controlled within 8 years.

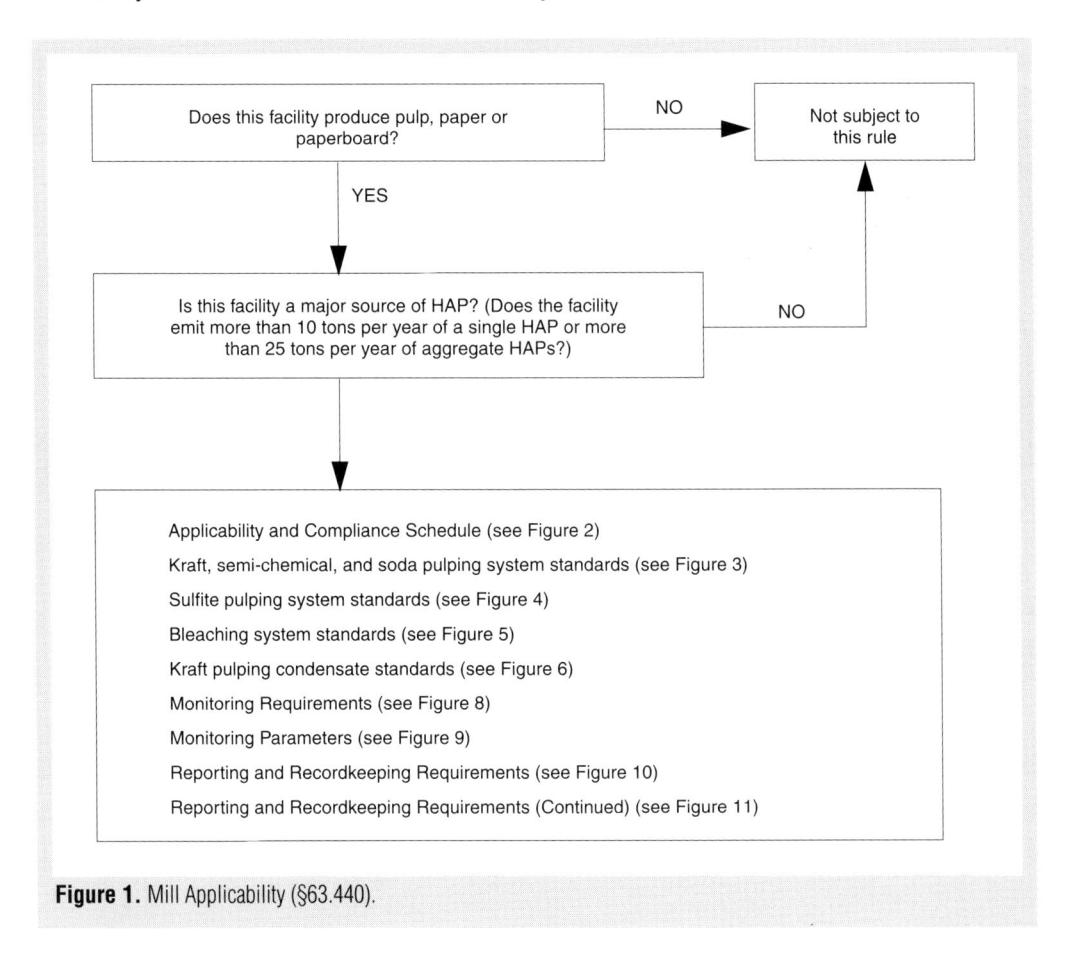

Figure 1. Mill Applicability (§63.440).

Appendix 1

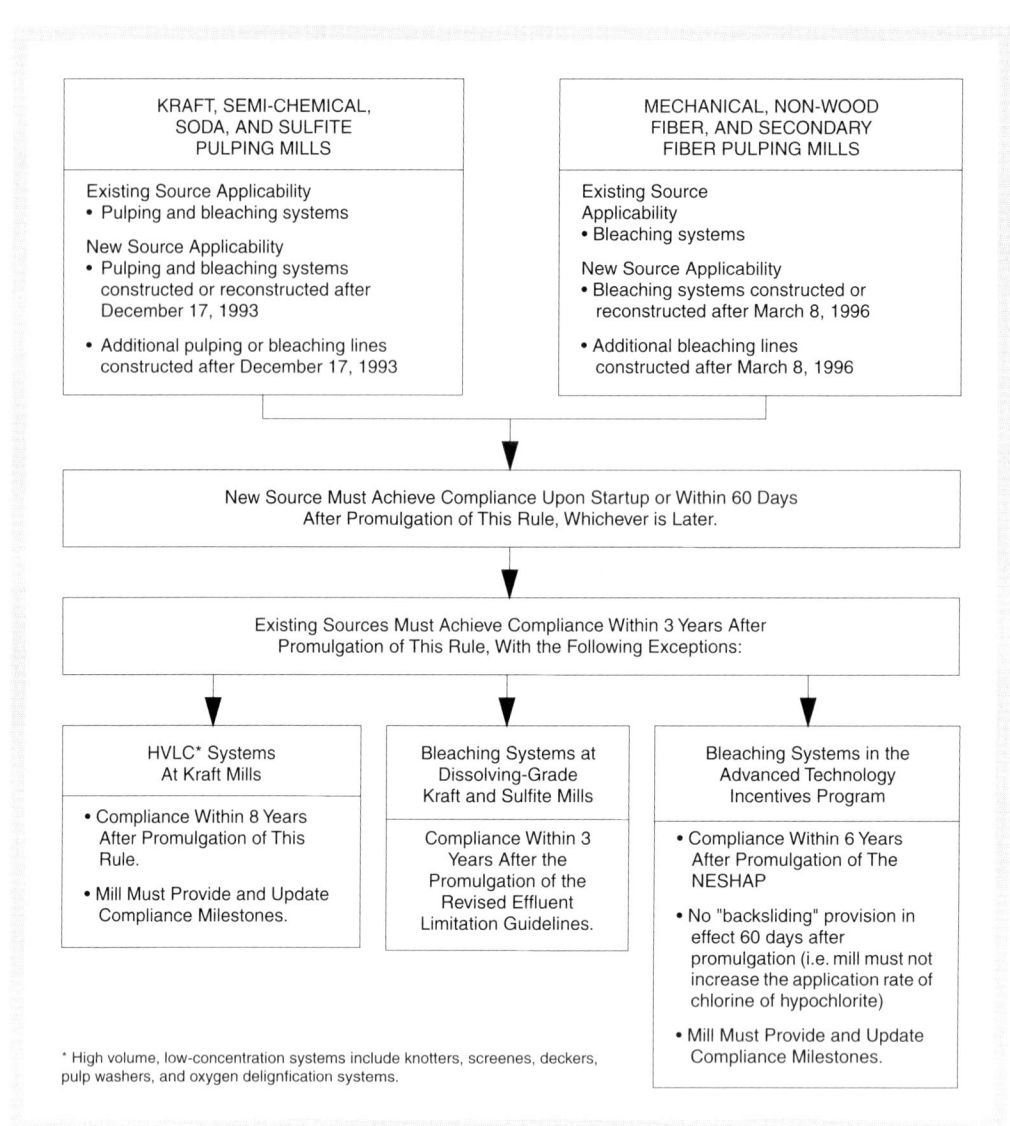

```
┌─────────────────────────────────────┐  ┌─────────────────────────────────────┐
│   KRAFT, SEMI-CHEMICAL,             │  │   MECHANICAL, NON-WOOD              │
│   SODA, AND SULFITE                 │  │   FIBER, AND SECONDARY              │
│   PULPING MILLS                     │  │   FIBER PULPING MILLS               │
│                                     │  │                                     │
│ Existing Source Applicability       │  │ Existing Source                     │
│ • Pulping and bleaching systems     │  │ Applicability                       │
│                                     │  │ • Bleaching systems                 │
│ New Source Applicability            │  │                                     │
│ • Pulping and bleaching systems     │  │ New Source Applicability            │
│   constructed or reconstructed after│  │ • Bleaching systems constructed or  │
│   December 17, 1993                 │  │   reconstructed after March 8, 1996 │
│                                     │  │                                     │
│ • Additional pulping or bleaching   │  │ • Additional bleaching lines        │
│   lines constructed after           │  │   constructed after March 8, 1996   │
│   December 17, 1993                 │  │                                     │
└─────────────────────────────────────┘  └─────────────────────────────────────┘
```

New Source Must Achieve Compliance Upon Startup or Within 60 Days After Promulgation of This Rule, Whichever is Later.

Existing Sources Must Achieve Compliance Within 3 Years After Promulgation of This Rule, With the Following Exceptions:

HVLC* Systems At Kraft Mills	Bleaching Systems at Dissolving-Grade Kraft and Sulfite Mills	Bleaching Systems in the Advanced Technology Incentives Program
• Compliance Within 8 Years After Promulgation of This Rule. • Mill Must Provide and Update Compliance Milestones.	Compliance Within 3 Years After the Promulgation of the Revised Effluent Limitation Guidelines.	• Compliance Within 6 Years After Promulgation of The NESHAP • No "backsliding" provision in effect 60 days after promulgation (i.e. mill must not increase the application rate of chlorine of hypochlorite) • Mill Must Provide and Update Compliance Milestones.

* High volume, low-concentration systems include knotters, screenes, deckers, pulp washers, and oxygen delignfication systems.

Figure 2. Applicability and Compliance Schedule (§63.440).

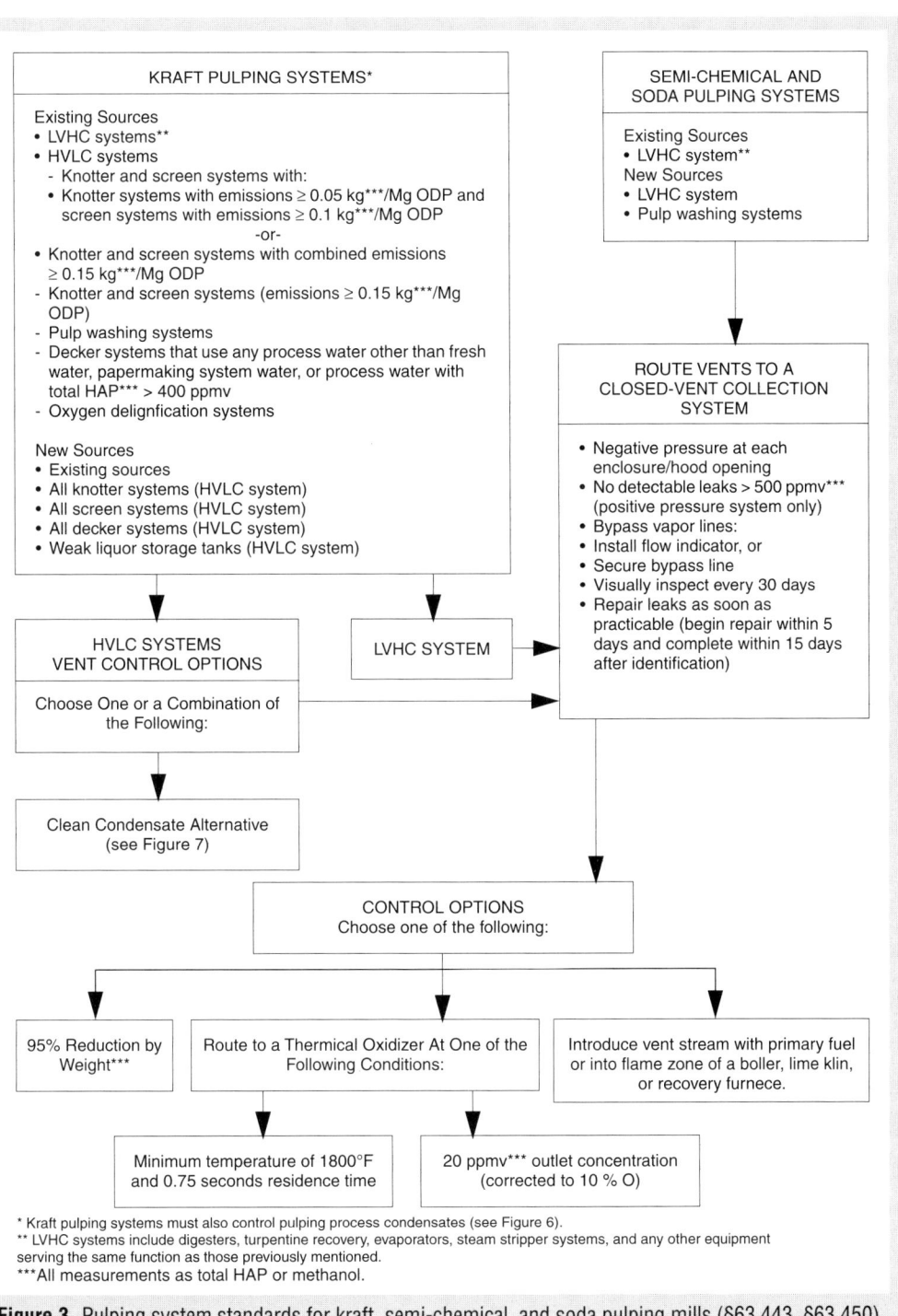

```
┌─────────────────────────────────────────────────┐   ┌──────────────────────────┐
│          KRAFT PULPING SYSTEMS*                 │   │   SEMI-CHEMICAL AND      │
│                                                 │   │   SODA PULPING SYSTEMS   │
│  Existing Sources                               │   │                          │
│  • LVHC systems**                               │   │  Existing Sources        │
│  • HVLC systems                                 │   │  • LVHC system**         │
│    - Knotter and screen systems with:           │   │  New Sources             │
│      • Knotter systems with emissions ≥ 0.05    │   │  • LVHC system           │
│        kg***/Mg ODP and screen systems with     │   │  • Pulp washing systems  │
│        emissions ≥ 0.1 kg***/Mg ODP             │   └────────────┬─────────────┘
│              -or-                               │                │
│      • Knotter and screen systems with combined │                │
│        emissions ≥ 0.15 kg***/Mg ODP            │                ▼
│    - Knotter and screen systems (emissions ≥    │   ┌──────────────────────────┐
│      0.15 kg***/Mg ODP)                         │   │   ROUTE VENTS TO A       │
│    - Pulp washing systems                       │   │   CLOSED-VENT COLLECTION │
│    - Decker systems that use any process water  │   │   SYSTEM                 │
│      other than fresh water, papermaking system │   │                          │
│      water, or process water with total         │   │ • Negative pressure at   │
│      HAP*** > 400 ppmv                          │   │   each enclosure/hood    │
│    - Oxygen delignfication systems              │   │   opening                │
│                                                 │   │ • No detectable leaks >  │
│  New Sources                                    │   │   500 ppmv***            │
│  • Existing sources                             │   │   (positive pressure     │
│  • All knotter systems (HVLC system)            │   │   system only)           │
│  • All screen systems (HVLC system)             │   │ • Bypass vapor lines:    │
│  • All decker systems (HVLC system)             │   │ • Install flow indicator,│
│  • Weak liquor storage tanks (HVLC system)      │   │   or                     │
│                                                 │   │ • Secure bypass line     │
└─────┬───────────────────────────┬───────────────┘   │ • Visually inspect every │
      │                           │                   │   30 days                │
      ▼                           ▼                   │ • Repair leaks as soon   │
┌──────────────────┐    ┌──────────────────┐          │   as practicable (begin  │
│  HVLC SYSTEMS    │    │   LVHC SYSTEM    │─────────▶│   repair within 5 days   │
│  VENT CONTROL    │    └──────────────────┘          │   and complete within 15 │
│  OPTIONS         │                                  │   days after             │
│                  │                                  │   identification)        │
│  Choose One or a │                                  └────────────┬─────────────┘
│  Combination of  │                                               │
│  the Following:  │                                               │
└────────┬─────────┘                                               │
         ▼                                                         │
┌──────────────────┐                                               │
│ Clean Condensate │                                               │
│ Alternative      │                                               │
│ (see Figure 7)   │                                               │
└──────────────────┘                                               │
                                                                   ▼
                          ┌──────────────────────────────────────────┐
                          │         CONTROL OPTIONS                  │
                          │      Choose one of the following:        │
                          └──┬────────────┬───────────────┬──────────┘
                             │            │               │
          ┌──────────────────┘            │               └──────────────────┐
          ▼                               ▼                                  ▼
┌──────────────────┐    ┌────────────────────────────┐    ┌─────────────────────────────┐
│ 95% Reduction by │    │ Route to a Thermal         │    │ Introduce vent stream with  │
│ Weight***        │    │ Oxidizer At One of the     │    │ primary fuel or into flame  │
│                  │    │ Following Conditions:      │    │ zone of a boiler, lime klin,│
└──────────────────┘    └──────┬───────────────┬─────┘    │ or recovery furnece.        │
                               ▼               ▼          └─────────────────────────────┘
                    ┌─────────────────┐ ┌────────────────────┐
                    │ Minimum         │ │ 20 ppmv*** outlet  │
                    │ temperature of  │ │ concentration      │
                    │ 1800°F and 0.75 │ │ (corrected to      │
                    │ seconds         │ │ 10 % O)            │
                    │ residence time  │ │                    │
                    └─────────────────┘ └────────────────────┘
```

* Kraft pulping systems must also control pulping process condensates (see Figure 6).
** LVHC systems include digesters, turpentine recovery, evaporators, steam stripper systems, and any other equipment serving the same function as those previously mentioned.
***All measurements as total HAP or methanol.

Figure 3. Pulping system standards for kraft, semi-chemical, and soda pulping mills (§63.443, §63.450).

Appendix 1

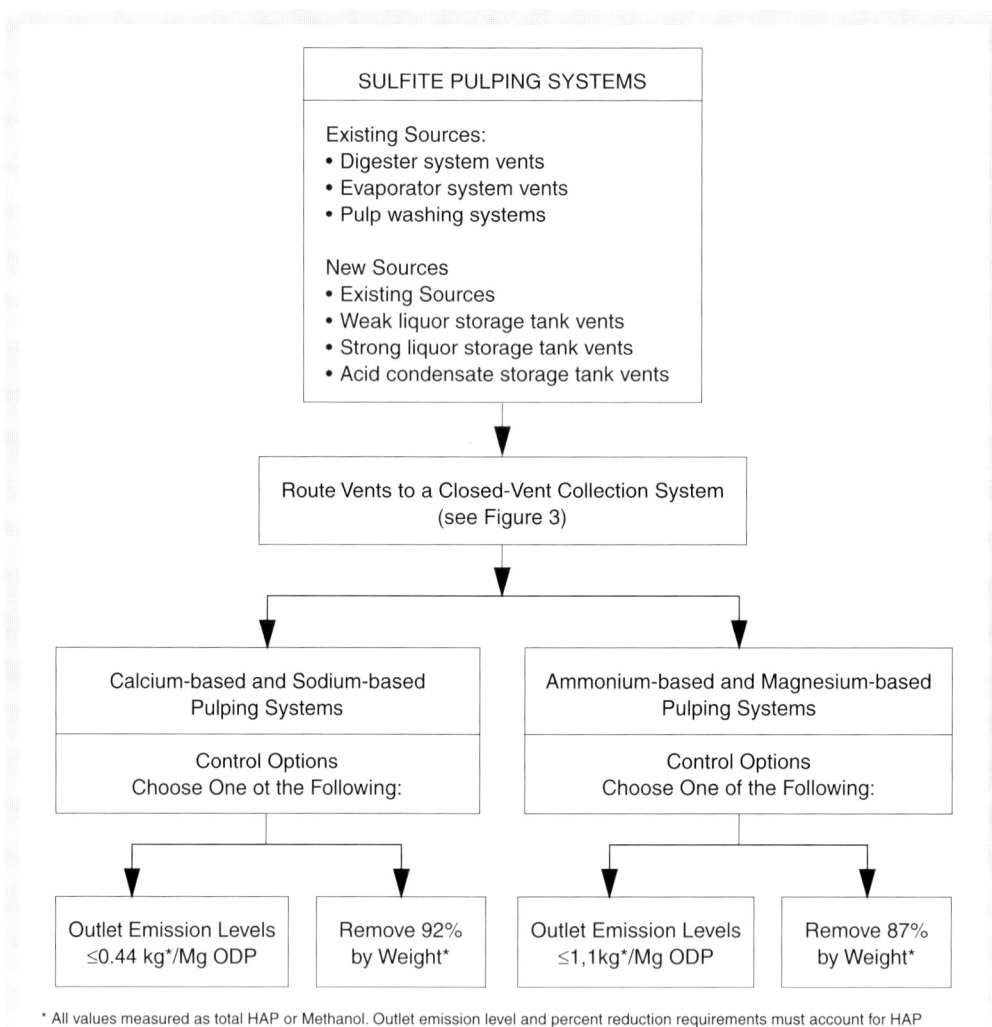

```
                    ┌─────────────────────────────────┐
                    │    SULFITE PULPING SYSTEMS      │
                    ├─────────────────────────────────┤
                    │ Existing Sources:               │
                    │  • Digester system vents        │
                    │  • Evaporator system vents      │
                    │  • Pulp washing systems         │
                    │                                 │
                    │ New Sources                     │
                    │  • Existing Sources             │
                    │  • Weak liquor storage tank vents│
                    │  • Strong liquor storage tank vents│
                    │  • Acid condensate storage tank vents│
                    └─────────────────────────────────┘
                                    │
                                    ▼
                  ┌──────────────────────────────────────┐
                  │ Route Vents to a Closed-Vent Collection│
                  │ System (see Figure 3)                │
                  └──────────────────────────────────────┘
```

Calcium-based and Sodium-based Pulping Systems — Control Options, Choose One of the Following:
- Outlet Emission Levels ≤0.44 kg*/Mg ODP
- Remove 92% by Weight*

Ammonium-based and Magnesium-based Pulping Systems — Control Options, Choose One of the Following:
- Outlet Emission Levels ≤1,1kg*/Mg ODP
- Remove 87% by Weight*

* All values measured as total HAP or Methanol. Outlet emission level and percent reduction requirements must account for HAP releases from vents, condensates, and wastewater from control devices used to reduce HAP emissions.

Figure 4. Pulping system standards for sulfite pulping mills (§63.443, §63,460).

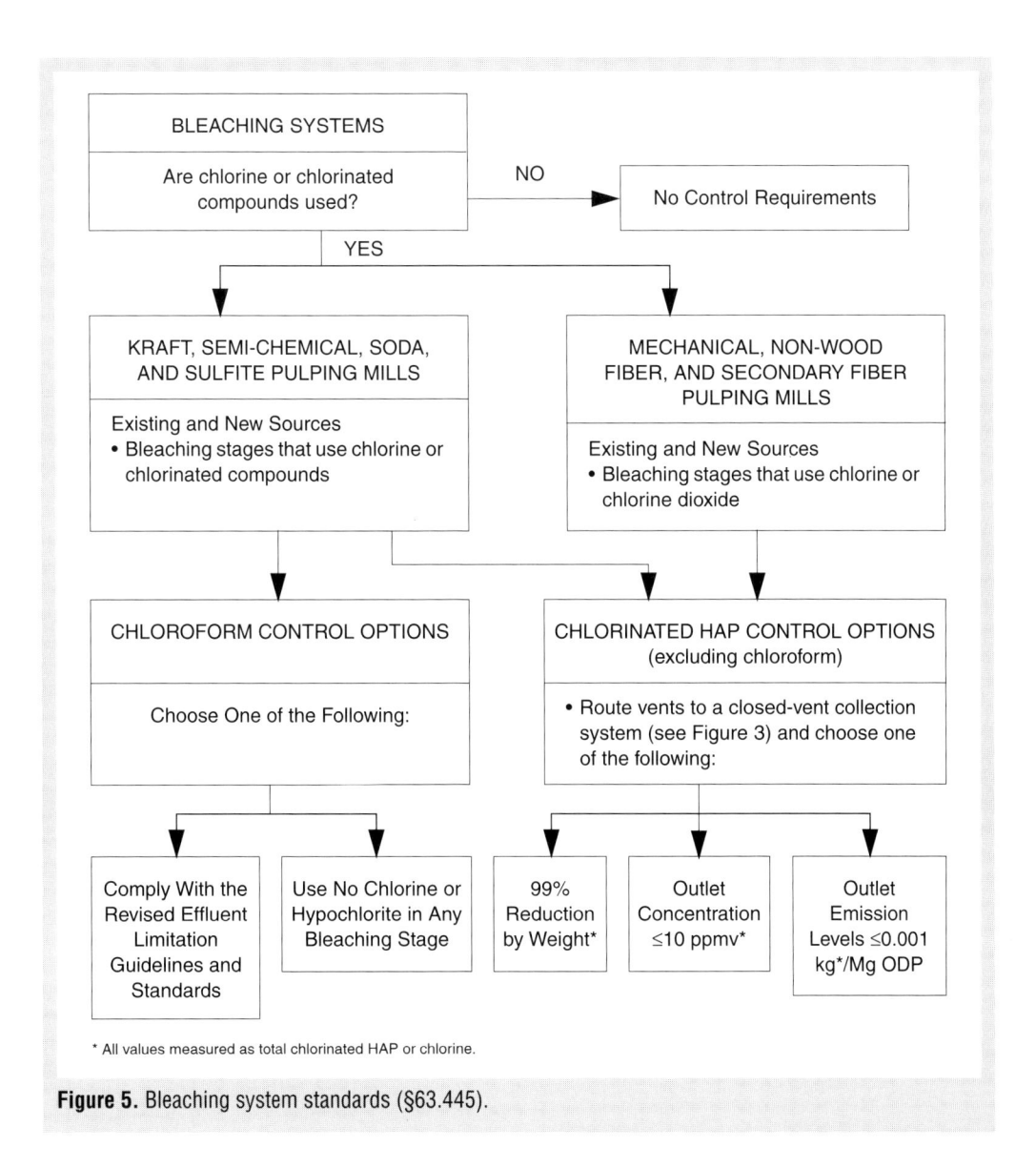

```
┌─────────────────────────────────┐
│      BLEACHING SYSTEMS          │
├─────────────────────────────────┤         NO      ┌──────────────────────────┐
│  Are chlorine or chlorinated    │───────────────▶│  No Control Requirements │
│     compounds used?             │                 └──────────────────────────┘
└─────────────────────────────────┘
              │ YES
    ┌─────────┴─────────┐
    ▼                   ▼
```

┌──────────────────────────────────────┐   ┌──────────────────────────────────────┐
│ KRAFT, SEMI-CHEMICAL, SODA,   │   │ MECHANICAL, NON-WOOD   │
│ AND SULFITE PULPING MILLS   │   │ FIBER, AND SECONDARY FIBER   │
├──────────────────────────────────────┤   │ PULPING MILLS   │
│ Existing and New Sources   │   ├──────────────────────────────────────┤
│ • Bleaching stages that use chlorine │   │ Existing and New Sources   │
│   or chlorinated compounds   │   │ • Bleaching stages that use chlorine │
│   │   │   or chlorine dioxide   │
└──────────────────────────────────────┘   └──────────────────────────────────────┘

┌──────────────────────────────────────┐   ┌──────────────────────────────────────┐
│   CHLOROFORM CONTROL OPTIONS   │   │ CHLORINATED HAP CONTROL OPTIONS   │
│   │   │   (excluding chloroform)   │
├──────────────────────────────────────┤   ├──────────────────────────────────────┤
│   Choose One of the Following: │   │ • Route vents to a closed-vent   │
│   │   │   collection system (see Figure 3)   │
│   │   │   and choose one of the following: │
└──────────────────────────────────────┘   └──────────────────────────────────────┘

| Comply With the Revised Effluent Limitation Guidelines and Standards | Use No Chlorine or Hypochlorite in Any Bleaching Stage | 99% Reduction by Weight* | Outlet Concentration ≤10 ppmv* | Outlet Emission Levels ≤0.001 kg*/Mg ODP |

* All values measured as total chlorinated HAP or chlorine.

Figure 5. Bleaching system standards (§63.445).

Appendix 1

```
                    ┌─────────────────────────────────────────────┐
                    │     KRAFT PULPING PROCESS CONDENSATES        │
                    │                                              │
                    │  New and Existing Sources:                   │
                    │  * Digester system                           │
                    │  * Turpertine recovery system                │
                    │  * Each weak liquor feed stage in the        │
                    │    evaporator system                         │
                    │  * LVHC collection system                    │
                    │  * HVLC collection system                    │
                    └─────────────────────────────────────────────┘
                                       │
                                       ▼
         ┌──────────────────────────────────────────────────────────┐
         │ Convey condensates in a closed collection system meeting │
         │ the requirements for individual drain systems as         │
         │ specified in Subpart RR (40 CFR 83.446 (d)) until the    │
         │ condensates reach one of the treatment options.          │
         └──────────────────────────────────────────────────────────┘
                                       │
                                       ▼
                         ┌──────────────────────────┐
                         │     CONTROL OPTIONS      │
                         │ Choose One of the Following: │
                         └──────────────────────────┘
```

CONTROL THE ENTIRE VOLUME OF THE CONDENSATES FROM ALL LISTED SOURCES

VOLUME REDUCTION OPTIONS (Condensate Segregation)

Segregate condensate streams into low-HAP and high-HAP streams. Choose one of the following:

VOLUME REDUCTION OPTION 1:

High-HAP stream that contains at least 85 percent of the total HAP* mass from the digester, turpentine recovery, and evaporator systems.

Control the high-HAP stream along with condensates from the LVHC and HVLC collection systems.

LOW-HAP STREAMS
* No further MACT requirements
* Return to mill or sower

VOLUME REDUCTION OPTION 2:

High-HAP stream that contains
> 3.6 kg*/Mg ODP (unbleached mills) or
> 5.5 kg*/Mg ODP (bleached mills)
from all listed sources.

Control the high-HAP stream.

TREATMENT OPTIONS
Choose One of the Following:

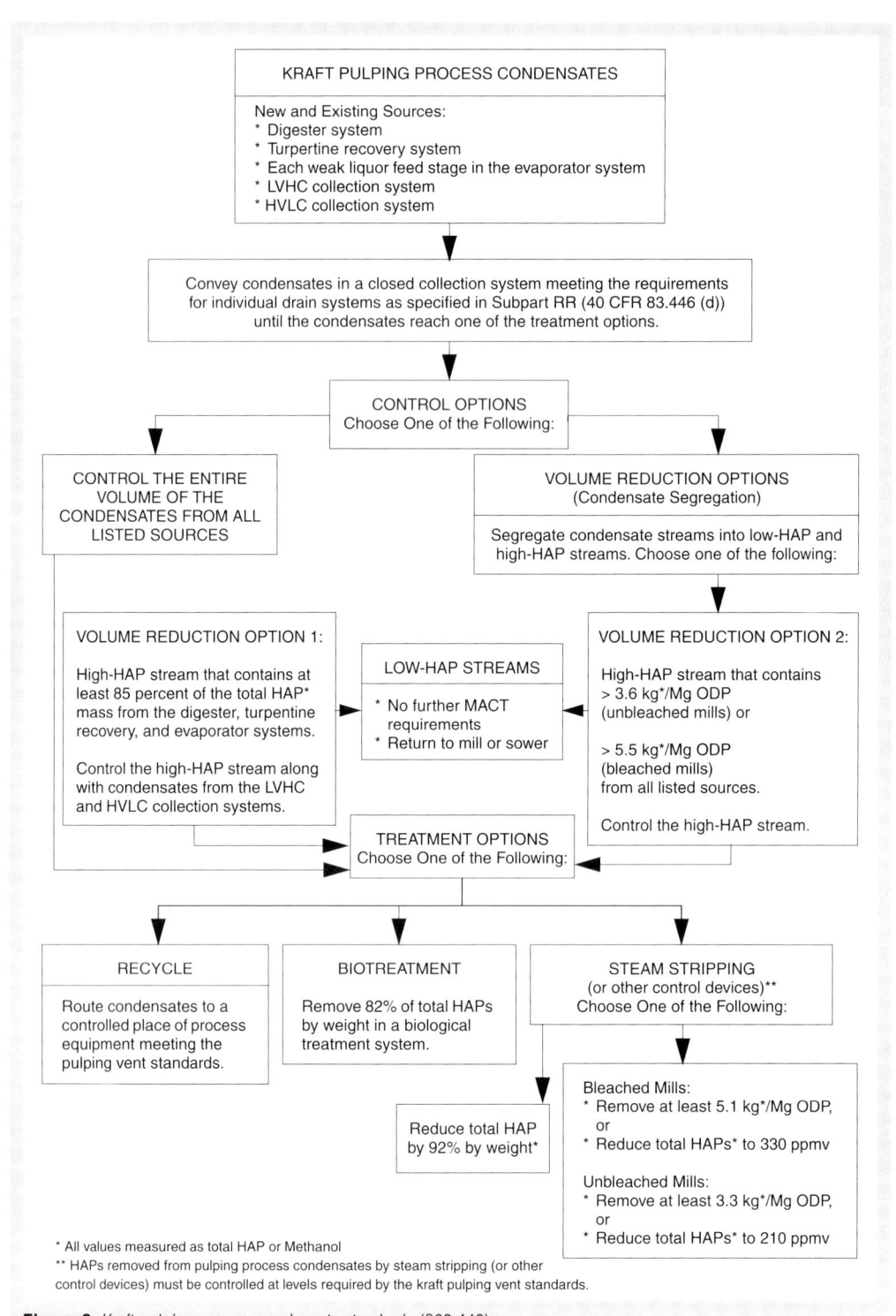

RECYCLE

Route condensates to a controlled place of process equipment meeting the pulping vent standards.

BIOTREATMENT

Remove 82% of total HAPs by weight in a biological treatment system.

Reduce total HAP by 92% by weight*

STEAM STRIPPING (or other control devices)**
Choose One of the Following:

Bleached Mills:
* Remove at least 5.1 kg*/Mg ODP, or
* Reduce total HAPs* to 330 ppmv

Unbleached Mills:
* Remove at least 3.3 kg*/Mg ODP, or
* Reduce total HAPs* to 210 ppmv

* All values measured as total HAP or Methanol
** HAPs removed from pulping process condensates by steam stripping (or other control devices) must be controlled at levels required by the kraft pulping vent standards.

Figure 6. Kraft pulping process condensate standards (§63.446).

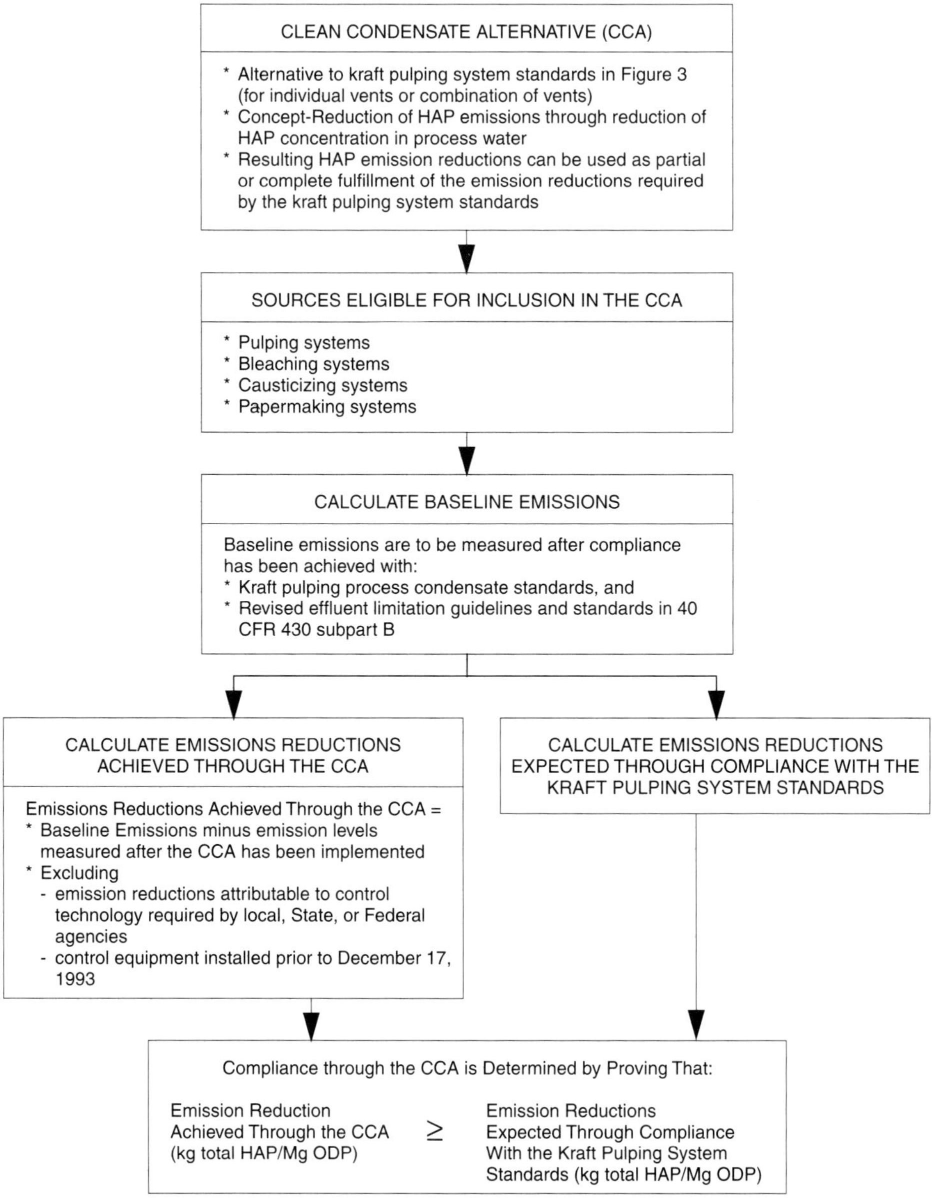

```
┌─────────────────────────────────────────────────────────┐
│              CLEAN CONDENSATE ALTERNATIVE (CCA)          │
│                                                          │
│  * Alternative to kraft pulping system standards in      │
│    Figure 3 (for individual vents or combination of vents)│
│  * Concept-Reduction of HAP emissions through reduction  │
│    of HAP concentration in process water                 │
│  * Resulting HAP emission reductions can be used as      │
│    partial or complete fulfillment of the emission       │
│    reductions required by the kraft pulping system       │
│    standards                                             │
└─────────────────────────────────────────────────────────┘
                          │
                          ▼
┌─────────────────────────────────────────────────────────┐
│           SOURCES ELIGIBLE FOR INCLUSION IN THE CCA      │
│                                                          │
│  * Pulping systems                                       │
│  * Bleaching systems                                     │
│  * Causticizing systems                                  │
│  * Papermaking systems                                   │
└─────────────────────────────────────────────────────────┘
                          │
                          ▼
┌─────────────────────────────────────────────────────────┐
│               CALCULATE BASELINE EMISSIONS               │
│                                                          │
│  Baseline emissions are to be measured after compliance  │
│  has been achieved with:                                 │
│  * Kraft pulping process condensate standards, and       │
│  * Revised effluent limitation guidelines and standards  │
│    in 40 CFR 430 subpart B                               │
└─────────────────────────────────────────────────────────┘
              │                              │
              ▼                              ▼
┌───────────────────────────┐   ┌───────────────────────────┐
│ CALCULATE EMISSIONS       │   │ CALCULATE EMISSIONS       │
│ REDUCTIONS ACHIEVED       │   │ REDUCTIONS EXPECTED       │
│ THROUGH THE CCA           │   │ THROUGH COMPLIANCE WITH   │
│                           │   │ THE KRAFT PULPING SYSTEM  │
│ Emissions Reductions      │   │ STANDARDS                 │
│ Achieved Through the CCA =│   └───────────────────────────┘
│ * Baseline Emissions minus│
│   emission levels measured│
│   after the CCA has been  │
│   implemented             │
│ * Excluding               │
│  - emission reductions    │
│    attributable to control│
│    technology required by │
│    local, State, or       │
│    Federal agencies       │
│  - control equipment      │
│    installed prior to     │
│    December 17, 1993      │
└───────────────────────────┘
              │                              │
              ▼                              ▼
┌─────────────────────────────────────────────────────────┐
│    Compliance through the CCA is Determined by Proving  │
│    That:                                                 │
│                                                          │
│    Emission Reduction            Emission Reductions     │
│    Achieved Through the CCA  ≥   Expected Through        │
│    (kg total HAP/Mg ODP)         Compliance With the     │
│                                  Kraft Pulping System    │
│                                  Standards (kg total     │
│                                  HAP/Mg ODP)             │
└─────────────────────────────────────────────────────────┘
```

Figure 7. Clean condensate alternative (§63.447).

Appendix 1

```
┌─────────────────────────────────────────┐
│         INSTALL AND OPERATE:            │
│    Continuous Monitoring System (CMS)   │
└─────────────────────────────────────────┘
           │                    │
           ▼                    ▼
```

ALLOWABLE OPERATING/EMISSION PARAMETERS FOR SPECIFIED SOURCES AND CONTROLS

* Use named parameters in Figure 9 for specified sources and controls
* Alternative parameters or processors to those specified are allowed only with EPA approval

DETERMINING OPERATING/EMISSION PARAMETERS FOR OTHER SOURCES AND CONTROLS

* Only for the sources and controls listed below:
  - Sulfite pulping vent controls.
  - Kraft condensate segregation techniques; and
  - Kraft clean condensate alternative controls.

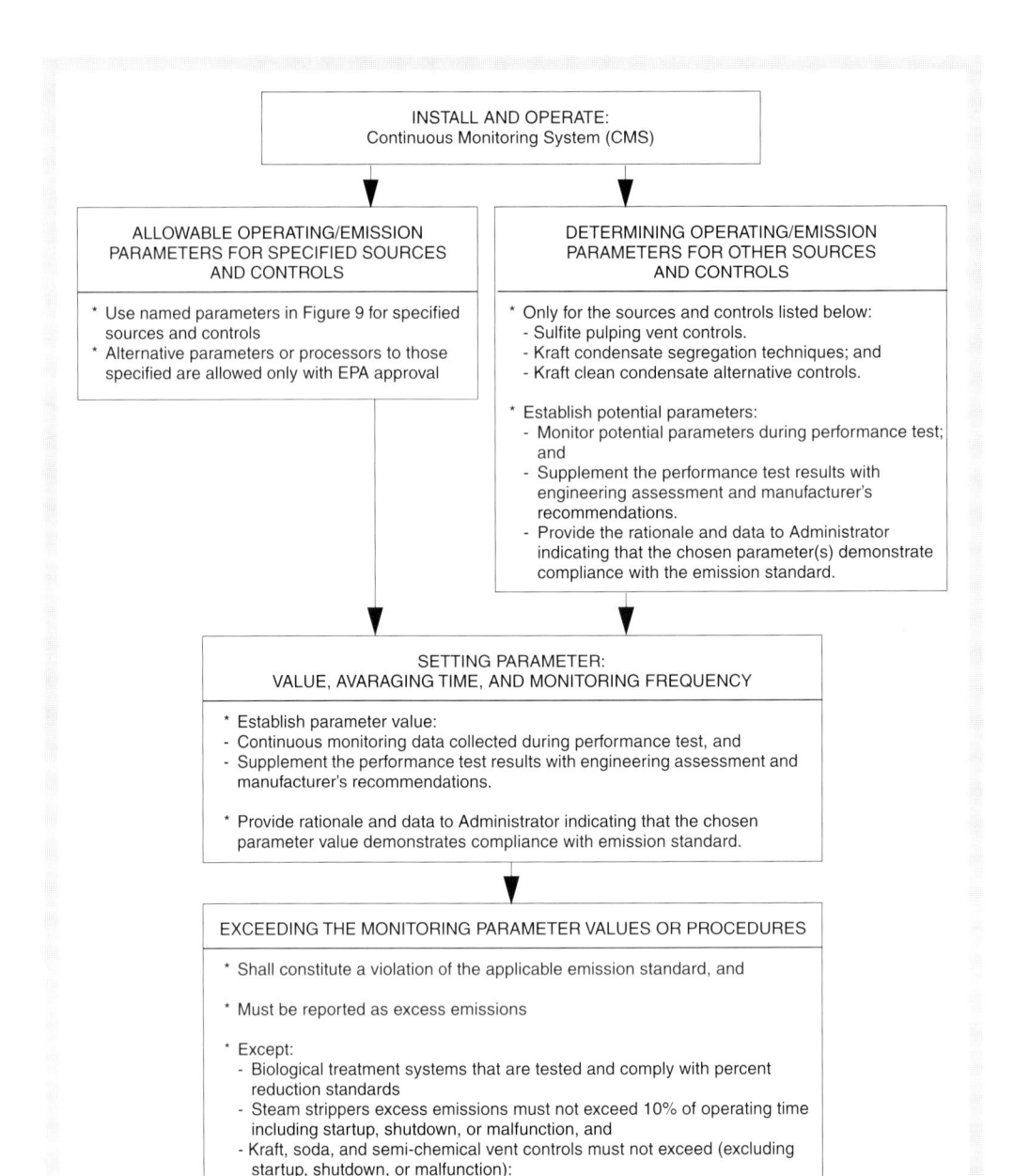

* Establish potential parameters:
  - Monitor potential parameters during performance test; and
  - Supplement the performance test results with engineering assessment and manufacturer's recommendations.
  - Provide the rationale and data to Administrator indicating that the chosen parameter(s) demonstrate compliance with the emission standard.

SETTING PARAMETER: VALUE, AVARAGING TIME, AND MONITORING FREQUENCY

* Establish parameter value:
  - Continuous monitoring data collected during performance test, and
  - Supplement the performance test results with engineering assessment and manufacturer's recommendations.

* Provide rationale and data to Administrator indicating that the chosen parameter value demonstrates compliance with emission standard.

EXCEEDING THE MONITORING PARAMETER VALUES OR PROCEDURES

* Shall constitute a violation of the applicable emission standard, and

* Must be reported as excess emissions

* Except:
  - Biological treatment systems that are tested and comply with percent reduction standards
  - Steam strippers excess emissions must not exceed 10% of operating time including startup, shutdown, or malfunction, and
  - Kraft, soda, and semi-chemical vent controls must not exceed (excluding startup, shutdown, or malfunction):
    * 1% for LVHC systems controls
    * 4% for HVLC or combined LVHC and HVLC system controls

Figure 8. Monitoring requirements (§63.453).

PULPING SYSTEMS	Thermal Oxidizer * For compliance with the 98 percent reduction option; measure, maintain, and record fire box temperature with a CMS* * For compliance with the 20 ppmv outlet option; measure, maintain, and record outlet HAP concentration with a CMS * For compliance with the 1800°F design temperature option; measure, maintain, and record fire box temperature with a CMS * No monitoring requirements for pulping vent system vents routed to a power boiler, lime klin, or recovery furnace

BLEACHING SYSTEMS	Bleaching Vent Scrubbers
	* Measure and record the following parameters using a CMS:
	  - pH or oxidation/reduction potential of scrubber effluent,
	  - gas scrubber inlet flowrate, and
	  - gas scrubber liquid influent flowrate
	  -or-
	  - Chlorine outlet concentration
	Systems participating in the extended compliance time of the Effluent Incentives Program:
	* Monitor chlorine and hypochlorite application rates (kg/Mg ODP) during extended compliance period

PULPING PROCESS CONDENSATES	Steam Strippers
	* Measure and record the following parameters using a CMS:
	  - Process water feed rate,
	  - Steam feed rate, and
	  - Column feed temperature
	  -or-
	  - Outlet methanol concentration
	Biological Treatment Systems
	* Daily monitoring
	  - Outlet soluble BOD
	  - Mixed liquor volatile suspended solids
	  - Horsepower of serator unit(s)
	  - Inlet liquid flow
	  - Liquid temperature
	  - Collect and store inlet and outlet grab samples
	* Quarterly Monitoring
	  - Every 1st quarter: demonstrate percent reduction of total HAP
	  - Remaining quarters: percent reduction of total HAP (methanol can be measured if a relationship between total HAP and methanol reduction is established and maintained at levels as less than those measured during 1st quarter

CLOSED VENT SYSTEMS -and- CLOSED (Condensates) COLLECTION SYSTEMS	Every 30 days: * Visual inspection * Inspecti bypass line valve or closure mechanism Initially and Annually * Demonstrate no detectable leaks at positive pressure portions * Demonstrate negative pressure at enclosure openings

* CMS = Continuous Monitoring System

Figure 9. Monitoring parameters (§63.453).

Appendix 1

INITIAL NOTIFICATION REPORT	NOTIFICATION OF COMPLIANCE STATUS REPORT
<u>Existing Major Sources:</u> Within 1 year after becoming subject to rule <u>New or Reconstructed Major Sources</u>: No later than 120 days after initial startup * Name & address of owner or operator. * Address of the source. * Identification of the rule and source's compliance date. * Description of operations, design capacity, and HAP emission points. * Statement of wheter a major or area source. * Notification of intent to construct or startup date for new or reconstructed sources. * Control strategy report (HVLC systems at kraft mills). * Control strategy report (bleaching systems participating in the Effluent Incentives Program).	60 Days Following Compliance Demonstration * Methods used to determine compliance. * Results or performance tests and/or CMS performance evaluations. * Methods to be used to determine continuous compliance. * Type and quantity of HAP emitted. * Analysis demonstrating wheter a major or area source. * Description of control equipment and efficiencies. * Statement as to wheter source has complied with standard. * Data, calculations, engineering assessments, and manufacturer's recommendations used to determine operating parameter value.

Figure 10. Recordkeeping and reporting requirements (§63.454, §63.455).

PERIODIC REPORTS	
QUARTERLY (Excess Emissions)	* Requirements specified in Subpart A * No additional requirements under Subpart S
SEMI-ANNUALLY (No Excess Emissions)	* Requirements specified in subpart A * Mills participating in the Effluent Incentives Program must report daily application rates of chlorine and hypochlorite
BI-ANNUALLY	* Mills with extended compliance schedules (some kraft pulping systems and for mills participating in the Effluent Incentives Program) must update control strategy reports

RECORDKEEPING
* Comply with recordkeeping requirements specified in Subpart A * Mills with closed-vent systems and/or closed collection systems shall prepare and maintain a site-specific inspection plan * Mills participating with the Effluent Incentives Program shall record daily avarage chlorine and hypochlorite application rates (kg/Mg ODP) * Mills shall record all CMS parameters included in the monitoring requirements (see Figures 8 and 9)

Figure 11. Recordkeeping and reporting requirements (§63.454, §63.455).

The EPA has set ambient air quality standards for the following pollutants: NOx, SO2, particulates, CO, O3, and Pb. Each state is in charge of maintaining the ambient air quality to fulfill the requirements as shown in Table 20, which also shows NSPS requirements for kraft mills.

Table 24. Ambient air quality restrictions and NSPS limits for kraft pulp mills in the United States (according to EPA).

	Ambient air	NSPS limits			
		Recovery boiler	Smelt dissolving	Lime kiln	Bark + oil boiler
Particulates	50 µg/m^3 (annual mean) DS	100 µg/m^{3}n 297 µg/m^3	0.15 kg/ metric tons	153 µg/ m^{3}n (gas) (oil)	0.043 g/MJ
TRS	N/A	6.7 µg/m^{3}n	0.017 kg/ metric tons DS	11.3 mg/ m^{3}n 0.34 g/MJ	
SO$_2$	80 µg/m^3 (annual mean)				
Nox	100 µg/m^3 (annual mean)				0.13 g/MJ
O$_3$	235 µg/m^3				

Each state is divided into various areas on the basis of the ambient air quality. So-called "attainment" areas, four classes in all, fulfill the ambient air quality criteria shown in Table 24. Other areas where the quality limits are exceeded are known as "nonattainment" areas. The states must improve the air quality in each nonattainment area over a five-year period from the date when the area has been classified as such by the EPA. This practice can cause wide variations in emission restrictions for similar mills depending on their location.

In 1992, the EPA required that, in those areas where the ozone limits are exceeded, VOC emissions be decreased by 15% by the end of 1996 and after that by 3% annually until the ozone concentration has decreased sufficiently. [Volatile Organic Compounds (VOC) can react with nitrogen oxides and produce photochemical oxidants, such as ozone.]

Mills located in attainment areas are subject to the Prevention of Significant Deterioration (PSD) principle; i.e., rebuild or increase of emission is not allowed without a permit procedure controlled by the EPA and the state in question.

Regulations on Waste Disposal

The goal of the EPA is to minimize waste formation, to reuse the material in waste, and/or to recover the energy value of the waste. The EPA has initiated several studies and programs to increase reuse of waste and decrease the amount of waste to be disposed of by landfill during the 1990s.

Appendix 1

With regard to hazardous waste, the EPA has issued regulations prohibiting certain types of wastes from entering municipal landfills. The wastes are specified as **listed wastes and characteristic wastes.**

Listed wastes consist of certain chemical compounds which have been deemed by the EPA to present significant risks to human health and the environment. In 1993, the list contained some 200 compounds, including the following heavy metals: antimony, arsenic, barium, cadmium, chromium, lead, mercury, and nickel. Organic compounds include methanol, phenols, benzenes, acetic acids, acetones, etc.

Characteristic wastes are wastes that are determined as being hazardous due to their nature. A waste is classified as a characteristic waste, if it is not a listed waste, but exhibits one or more of the following four characteristics:

- Ignitable (flash point < 1400 F)
- Corrosive (pH < 2 or pH > 12.5)
- Reactive (unstable, reacts with water)
- TCLP toxic (Toxicity Characteristic Leaching Procedure).

The major part of solid waste streams in pulp and paper mills can be classified as nonhazardous and can be disposed of as landfill. In the future, if closed-cycle operations are widely used, the properties of solid waste may change considerably.

Environmental Impact Assessment (EIA)

For new industrial enterprises for which NSPS have been developed, an environmental impact assessment must be carried out prior to the granting of a construction permit. The requirements of this assessment can be met by one of three options:

1) By categorical exclusion. This can be applied to cases which individually and cumulatively have no significant effect on the environment and no further analysis is required.

2) By preparing an Environmental Assessment (EA) document. This document outlines the environmental impact to determine any significant environmental effects

3) By preparing an Environmental Impact Statement (EIS). The EIS is to be made if at the EA stage significant environmental impact can be expected from the proposed activity. In an EIS, the direct and indirect environmental impacts of various alternatives for production, processing, and other activities are compared. The EPA can decide which, if any, of the alternatives is to be preferred on the basis of the results of the EIS.

Executive Authorities

The United States EPA is the primary regulatory authority. However, most of the federal legislation encourages individual states to assume the responsibility for establishing and enforcing regulations by providing federal funding for such programs. Most states have their own programs for air and water pollution control, and some also have solid and hazardous waste programs.

Intermediate between the federal and state authorities are the EPA regional authorities, which oversee environmental quality in ten regions of the United States, each comprised of several states.

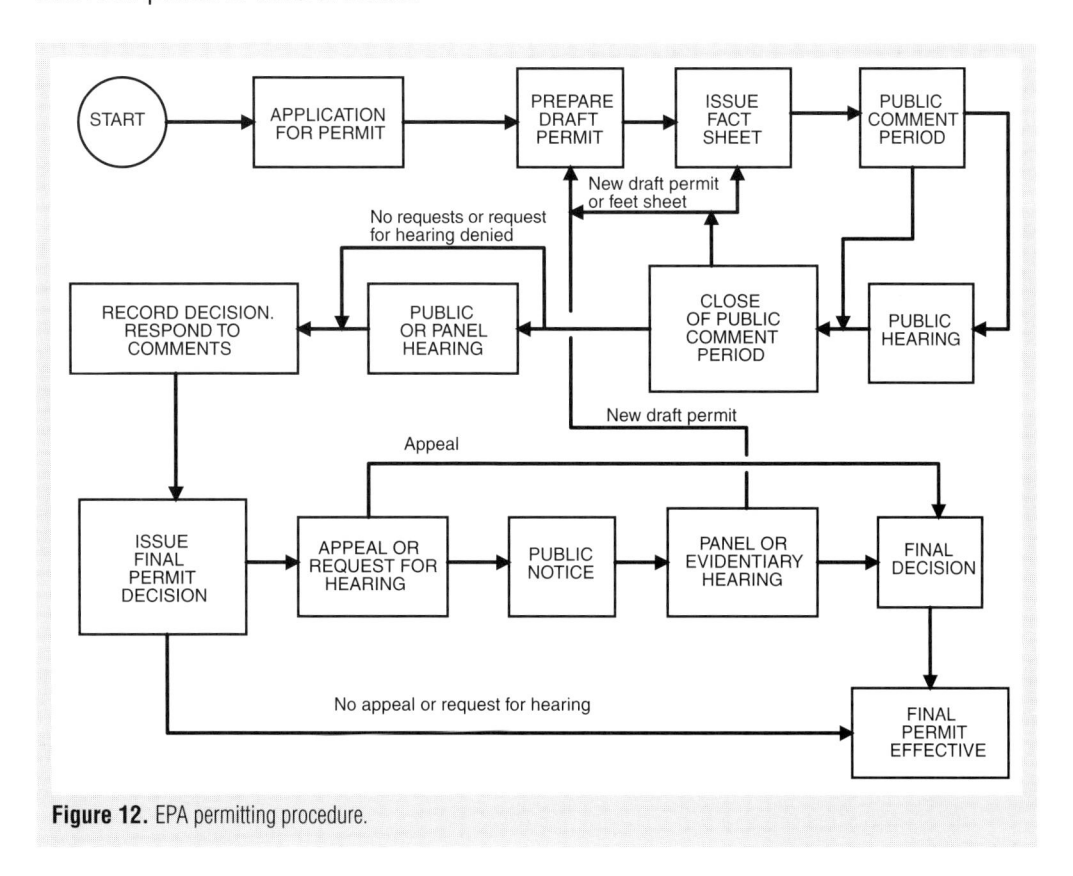

Figure 12. EPA permitting procedure.

Fig. 12 shows a typical permit procedure for an EPA-administered program. State programs generally follow the same procedure. As can be seen in the block diagram, several stages include public access to the procedure.

1.13 Canada

General Environmental Legislation

Both provincial and federal statutes govern environmental legislation in Canada. Provinces exercise proprietary rights over the resources (water, air) and therefore have the right to legislate on all aspects of water and air use and contamination.

Appendix 1

The federal body for administering environmental matters is Environment Canada. Other federal departments may be involved, such as the Department of Fisheries and Oceans and the Department of Public Health and Welfare.

Each province has its own provincial environmental body responsible for provincial legislation and implementation of laws.

The first environmental framework law for the pulp and paper industry was the Fisheries Act 1971, which promulgated certain effluent discharge regulations (BOD_5, TSS, and toxicity). In 1988, the Canadian Environmental Protection Act (CEPA) was adopted covering the regulation of toxic substances, etc.

The Department of Fisheries and Oceans revised the Federal Pulp and Paper Effluent Regulations in May 1992. The new regulations which apply to existing mills required compliance by Dec. 31, 1995. These regulations define generic emission limits for BOD, TSS, and toxicity. It should be noted that AOX was not included in the final regulations.

The Federal Government has set guidelines for ambient air quality standards. Most provinces follow these federal guidelines fairly closely. Several provinces have established ambient air standards for TRS compounds, but federal standards are lacking.

Presently there are no federal standards in force for air emissions from the pulp and paper industry. General guidelines were issued in 1979, but they are not incorporated in the CEPA laws.

Regulations on Effluent Discharges

The federal regulations 1992 define the maximum BOD and TSS discharge levels for pulp and paper mills. Acute toxicity of the effluent is regulated by the Federal Government.

A "Reference Production Rate" (RPR) is used to calculate daily or monthly allowable BOD and TSS discharges. The RPR is defined as the daily production of the finished product that the mill has been able to sustain 10 percent of the time. Furthermore, RPR is defined as the highest production capability over the past three years of operation. The maximum BOD and TSS are calculated according to the following equations:

1) Daily Discharge = $F \times 2.5 \times RPR$

2) Monthly Discharge = $F \times D \times 1.5 \times RPR$

where F is multiplier factor;
              for **calculation of BOD load**
                  18 for dissolved sulfite production
                  5 for all other pulp and paper production
              for **calculation of TSS load**
                  25 for dissolved sulfite production
                  7.5 for all other pulp and paper production
          D number of days in a month
          RPR Reference Production Rate.

Table 25 summarizes the current wastewater discharge limits according to the Federal regulations.

Table 25. Effluent discharge limits (expressed as kg/a.d. metric ton) by Federal Regulations in Canada.

Mill type	BOD		TSS	
	Daily	Monthly	Daily	Monthly
Dissolving sulfite mills	45.0	27.0	62.5	37.5
Other pulp and paper mills	12.5	7.5	18.75	11.25
	(24.75)[a]	(14.85)[a]		

[a]Maximum BOD that can be authorized for old mills, built before 1970

The provinces may set their own limits, equal to or more stringent than the federal regulations. In Alberta, there are no specific provincial regulations for the pulp and paper industry. Emission limits are determined for each mill individually. Emission limit development may be both technology based and ambient air and water quality based. In Alberta, the present effluent limit values are more stringent than the federal regulations. In British Columbia, Ontario, and Quebec the effluent restrictions are equal to or slightly tighter than the federal regulations.

Regulations on Atmospheric Emissions

Federal ambient air quality standards exist for SO_2, particulates, NO_x, CO, and ozone. Additionally, the provinces have issued their own ambient air standards, which follow quite closely those of federal requirements. Some provinces also have restrictions for TRS compounds.

Air emission standards exist also as federal guidelines and at provincial level for SO_2, particulates, and TRS.

Regulations on Waste Disposal

No federal regulations for waste disposal have been promulgated. Provinces have their own practice, e.g., special licenses for solid waste disposal or waste disposal permits are granted with the wastewater discharge permit.

Environmental Impact Assessment (EIA)

EIA procedure resembles the relevant US practice and, in most cases, it is obligatory to undertake EIA when, for example, a greenfield pulp and paper mill is built or a major rebuild is implemented.

Appendix 1

Executive Authorities

The executive authorities act at both federal and provincial level. At the federal level, the Ministry of the Environment has the primary responsibility concerning environmental administration. Other Ministries are also involved (e.g., The Ministry of Fisheries and Oceans, the Ministry of Health and Welfare).

The provinces are in charge of permit procedures, EIAs, monitoring, and follow-up of programs and mill operation.

1.14 International Conventions

Major multinational environmental conventions are briefly outlined. International conventions have been signed to protect extensive water areas and the atmosphere beyond national boundaries and to achieve a homogenous goal in the level of environmental protection between all parties. Still, it must be emphasized that in most cases, the agreements include only recommendations for decreasing or limiting pollution and thus the responsibility to implement the proposed measures lies with each individual party.

Water Pollution Control

The PARCOM Convention

The initiative in endeavoring to reach an agreement was taken by France in 1973. The convention was signed in June 4, 1974, in Paris (PARCOM). The convention was amended in 1986 by issuing the Paris Protocol.

The purpose of PARCOM is to protect marine pollution from land-based sources. The parties are obliged to undertake the following pollution control measures:

- **Eliminate**, when necessary by stages, the release of substances specified in Annex A (e.g., organohalogen compounds, Hg and its compounds, Cd and its compounds)

- **Strictly limit** the release of substances, such as organic compounds of P, Si and Sn, and elemental P, As, Cr, Co, Pb, Ni, Zn.

The PARCOM convention also includes various regulations and recommendations. In 1987, the Commission emphasized that all PARCOM decisions are legally obligatory and should be implemented in the national legislation of each of the signatories.

The Convention for the Protection of the Marine Environment of the North East Atlantic (OSPAR)

The OSPAR Convention was signed on Sept. 22, 1992, in Paris. OSPAR is intended to replace the PARCOM Convention after its ratification among the signatories.

The OSPAR Convention emphasizes especially the precautionary and the polluter pays principles of Community environmental policy. OSPAR includes the concepts of Best Available Techniques (BAT) and Best Environmental Practice (BEP)[4].

BAT is defined as the latest stage of development of processes, facilities, or methods of operation which indicate the practical suitability of a particular measure for limiting discharges, emissions, and waste. Special consideration in BAT evaluations should be paid to:

- Presenting comparable processes, facilities, or methods of operation which have recently been successfully tried out
- Technological advances and changes in scientific knowledge and understanding
- The economic feasibility of such techniques
- Time limits for installation in both new and existing plants
- The nature and volume of the discharges and emissions concerned.

BEP is defined as the application of the most appropriate combination of environmental control measures and strategies. BEP presupposes that at least the following measures should be considered:

- The provision of information and education to the public and to users about the environmental consequences of choosing particular activities and products, and of the use and disposal of these products
- The development and application of codes of good environmental practice which cover all aspects of the activity in the product's life
- The establishment of a system of licensing, involving a range of restrictions or a ban.

According to Annex I of the convention, the use of BAT will be applied to point sources and BEP for point and diffuse sources.

The Helsinki Convention (HELCOM)

The Convention on the protection of the marine environment of the Baltic Sea area was established at Finland's initiative in Helsinki in March 1974.

The current amendment of the Convention, which was signed on April 9, 1992, also includes the definitions of BAT and BEP. The parties agree to promote the use of BAT and BEP. BAT is to be used for point sources of pollution and BEP for all sources of pollution.

HELCOM 1992 also includes regulations on issuing permits for industrial plants (pollution from land-based sources). The appropriate national authority shall issue the permit after comprehensive assessment with special consideration of the above mentioned principles. Minimum requirements for each permit are stated as follows:

- Limit values for amount and quality (load and/or concentration) of direct and indirect discharges and emissions
- Type and extent of control to be performed by the operator (self control) and analytical methods to be used.

Appendix 1

The appropriate national authority or an independent authorized institution shall inspect the amount and quality of discharges and/or emissions by sampling and analyzing.

With regard to the pulp and paper industry, the following HELCOM recommendations are adopted as revised in March 1996:

- HELCOM recommendation 17/8 (reduction of discharges from the kraft pulp industry)
- HELCOM recommendation 17/9 (reduction of discharges from the sulfite pulp industry.

Table 26 shows discharge limits according to HELCOM 17/8.

Table 26. Annual average discharge limit values for the kraft pulp industry according to HELCOM recommendation 17/8.

The following annual average discharge limit values in kg per metric ton of Air Dry Pulp (kg/t ADP) produced are not exceeded from Jan. 1, 2000, for any mill which has started to operate before Jan. 1, 1997:				
Pulping process	COD_{Cr}	AOX	Tot-P	Tot-N
Bleached pulp	30	0.4	0.04	0.4
Unbleached pulp	15	–	0.02	0.3
In countries in transition, the following annual average discharge limit values (kg/t ADP) produced are not exceeded from Jan. 1, 2005, for any mill which has started to operate before Jan. 1, 1997:				
Pulping process	COD_{Cr}	AOX	Tot-P	Tot-N
Bleached pulp	35	0.4	0.04	0.4
Unbleached pulp	20	–	0.02	0.3
For any mill, starting to operate or considerably increasing its capacity (by more than 50%) after Jan. 1, 1997, the following annual discharge limit values (kg/t ADP) exist:				
Pulping process	COD_{Cr}	AOX	Tot-P	Tot-N
Bleached pulp	15	0.2	0.02	0.35
Unbleached pulp	8	–	0.01	0.25

In addition, **HELCOM 17/8** also recommends that:

- Molecular chlorine not be used in the bleaching of kraft pulp after Jan. 1, 1997 (2000 for countries in transition)
- Limit values for nitrogen should apply to kraft pulp mills located on the coast
- The signatories should report every three years starting in 2000
- With the development of BAT, especially with the change of chelating agents to biodegradable compounds, the recommendation must be reconsidered in 1998.

In Attachment 1, HELCOM 17/8 determines BAT for the kraft pulp industry, 1995 as follows:

1. Dry debarking with minor wastewater discharges
2. Closed screening
3. Stripping of most concentrated condensates and reuse of most condensates in the process
4. Systems which enable the recovery of almost all spillages
5. Extended delignification in the digester followed by oxygen delignification
6. Efficient washing before the pulp leaves the closed part of the process
7. At least secondary treatment for wastewater discharges
8. Partial closure of bleach plants. The main part of the discharge from bleach plants is to be piped to the recovery system.
9. Use of environmentally sound chemicals in the process, for example, use of biodegradable chelating agents wherever possible.

Attachment 2 gives analysing methods to be applied for AOX, COD_{Cr}, Tot-P, and Tot-N (all analyses to be made on unsettled, unfiltered samples).

A recommendation for reporting procedure is also given.

Table 27 shows discharge limits according to **HELCOM 17/9**.

Table 27. Annual average discharge limit values for the sulfite pulp industry according to HELCOM recommendation 17/9.

The following annaual average discharge limit values (kg/metric ton of a.d. pulp) produced are not exceeded from Jan. 1, 2000 (for countries in transition from Jan. 1, 2005) for any mill which has started to operate before Jan. 1, 1997:				
Pulping process	COD_{Cr}	**AOX**	**Tot-P**	**Tot-N**
Bleached pulp	70	0.5	0.08	0.7
Unbleached pulp	45	–	0.06	0.6
For any mill, starting to operate or considerably increasing its capacity (by more than 50%) after Jan. 1, 1997, the following annual discharge limit values (kg/metric ton of a.d. pulp) exist:				
Pulping process	COD_{Cr}	**AOX**	**Tot-P**	**Tot-N**
Bleached pulp	35	0.1	0.04	0.4
Unbleached pulp	20	–	0.03	0.3

Additional recommendations in HELCOM 17/9 are:

- Molecular chlorine is not to be used in the bleaching of sulfite pulp after Jan. 1, 1997.
- Limit values for nitrogen should apply to sulfite mills located on the coast.

Appendix 1

- The signatories should report to the Commission every three years starting in 2000.
- With the development of BAT and the use of chelating agents, the recommendation should be reconsidered in 1998.

In Attachment 1 of HELCOM 17/9, BAT for the sulfite industry 1995 is determined as follows:

1. Dry debarking with minor wastewater discharges
2. Closed screening
3. Neutralizing of weak liquor before evaporation followed by reuse of the main part of condensates in the process
4. Systems which enable the recovery of almost all organic substances dissolved in the cook
5. No discharge from bleach plants when sodium-based processes are being used
6. At least secondary treatment for wastewater discharges
7. Partial closure of bleach plants when processes other than sodium-based ones are used
8. Use of environmentally sound chemicals in the process, such as the use of biodegradable chelating agents wherever possible.

Attachment 2 contains similar analysis and monitoring recommendation to HELCOM 17/8.

Proposals of the Nordic Council of Ministers

The Nordic Council of Ministers established a working group comprising experts from Sweden, Finland, Norway, and Denmark to evaluate the impact of the pulp and paper industry on the environment and to compile a report on it by 1993. The Nordic Council of Ministers (Study on Nordic Pulp and Paper Industry and the Environment) issued the report on Nov. 7, 1993.

This report presented the following proposals related to environmental protection and pollution control[5]:

1. The Nordic Pulp and Paper Industry should strive toward the use of cleaner production technology.
2. Molecular chlorine should not be used when bleaching chemical pulp after Jan. 1, 1996 (reservations by pulp industry representatives indicated that a ban on the use of molecular chlorine cannot be accepted, even though the use of chlorine gas will most probably cease after 1995).

3. Pollution loads, as shown in Table 28, expressed as annual averages, should not be exceeded for **any mill** by the end of this century. For integrated mills producing mechanical or recycled fibers pulp, the figures are given as kg per metric ton of product, whereas for other types of mills the figures are given as kg per metric ton of air dry pulp.

Table 28. Annual average limit values of pollution loads (kg/a.d. metric ton) according to a proposal by the Nordic Council of Ministers 1993. Limit values to be reached by the end of this century.

Type of mill	AOX	COD$_{Cr}$	Tot-P	Tot-N[a]	Sulfur[b]	NO$_x$[c]
Bleached kraft	0.4	30	0.04	0.2	1.0	1.5
Unbleached kraft	–	15	0.02	0.2	1.0	1.5
Bleached sulfite	0.3	70	0.08	0.6	1.5	2.0
CTMP	–	–	30	0.02	0.2	–
Mechanical[d]	–	10	0.01	0.2	–	–
Recycled fiber	–	10	0.01	0.2	–	–

[a]Any nitrogen discharge associated with the use of complexing agents should be added to the figure for tot-N given above
[b]Gaseous sulphur emissions, as S, except from auxiliary boilers
[c]Gaseous nitrogen oxide emissions, as NO$_2$, except from auxiliary boilers
[d]"Mechanical" means integrated mills producing newsprint or magazine paper

4. In the case of any **new or considerably enlarged** (on the order of 30%) mill, the following levels, as shown in Table 29, should not be exceeded as annual averages.

Table 29. Annual average limit values of pollution loads (kd/a.d metric ton) according to a proposal by the Nordic Council of Ministers 1993. New and enlarged mills.

Type of mill	AOX	COD$_{Cr}$	Tot-P	Tot-N[a]	Sulfur[b]	NO$_x$[c]
Bleached kraft	0.2	15	0.02	0.15	0.5	1.0
Unbleached kraft	–	8	0.01	0.15	0.5	1.0
Bleached sulfite	0.1	35	0.04	0.3	1.0	1.0
CTMP	–	15	0.01	0.1	–	–
Mechanical[d]	–	5	0.005	0.1	–	–
Recycled fiber	–	5	0.005	0.1	–	–

[a]Any nitrogen discharge associated with the use of complexing agents should be added to the figure for tot-N given above.
[b]Gaseous sulfur emissions, as S, except from auxiliary boilers
[c]Gaseous nitrogen oxide emissions, as NO$_2$, except from auxiliary boilers
[d]"Mechanical" means integrated mills producing newsprint or magazine paper

Appendix 1

5. The Nordic Council of Ministers should encourage the Nordic countries and their industry to support:

    a) Studies on the effects of discharges of metals from the pulp and paper industry

    b) The development of treatment methods which can remove complexing agents like EDTA and further support research on possible environmental effects of such substances

    c) Research on the environmental effects of both current and new technologies used in the pulp and paper industry.

6. The Nordic mills should have implemented a waste minimization program within a period of five years.

7. The environmental authorities of the Nordic countries should exchange experience on methods of sampling and analyses of inputs with a view to harmonization.

8. The environmental authorities of the Nordic countries should exchange experience on monitoring.

9. The environmental authorities of the Nordic countries should exchange information on the hazards of chemicals used in the Nordic pulp and paper industry and possible alternatives to substances regarded as hazardous to the environment.

10. The environmental authorities of the Nordic countries should exchange reports annually on input from the pulp and paper industry.

11. The environmental authorities of the Nordic countries should cooperate in evaluating the environmental impact of both current and new technologies used in the pulp and paper industry. Moreover, the countries should promote research in this field.

12. The environmental authorities of the Nordic countries should encourage the pulp and paper industry to introduce and support Environmental Management Systems, including Environmental Auditing.

Air Pollution Control

The 1979 Geneva Convention (Convention on Long-range Transboundary Air Pollution)

The Geneva Convention is the cornerstone of international sulfur and nitrogen emission policy. The following protocols have been promulgated on the basis of the Geneva Convention:

- The Geneva Protocol 1984 (monitoring and evaluation)
- The Helsinki Protocol 1985 (sulfur compounds)
- The Sofia Protocol 1988 (nitrogen oxides).

The Protocols include goals for future air emissions and claim to use BAT to achieve these targets. Development and harmonization of the monitoring and analysing procedures between the contracting parties is also required.

The 1985 Vienna Convention

The Vienna Convention was signed for the protection of the ozone layer and was completed by the Montreal Protocol in 1987. It contains a list of specially controlled substances which deplete the ozone layer.

The 1992 Framework Convention on Climate Change (Rio de Janeiro)

The Framework Convention by the United Nations has been ratified in some 160 UN member states. All EU member countries have ratified the Convention. The purpose of the 1992 Rio Convention is to stabilize the content of greenhouse gases in the atmosphere at a level where no harmful or dangerous changes can occur. No numerical target concentrations, however, are given.

In 1995, the contracting parties began negotiations to adopt a binding document to be associated with the Convention (the Berlin Mandate). The document is planned to include the measures necessary in practice to decrease the emissions of greenhouse gases (also regional quantitative restrictions) and to protect the carbon resources in nature. According to the plans, the negotiations will be completed during 1997.

References

1. Jans, J.H., *European Environmental Law*, Kluwer Law International, The Hauge-London-Boston, 1995, pp. 1–388.

2. *The Use of Economic Instruments in Nordic Environmental Policy*, Nordic Council of Ministers/TemaNord Environment, 1996:568, Stockholm, p. 114.

3. *Skogsindustrins utsläpp till vatten & luft samt avfallsmängder 1995*, Naturvårdsverket, Stockholm, 1993, p. 81.

4. *Environmental Emission Data – International Comparability*, Finnish Environment Agency, series A 221, Helsinki, 1995, p. 48.

5. *Study on Nordic Pulp and Paper Industry and the Environment*, Nordic Council of Ministers, 1993:638, Stockholm, 1993, p. 81.

6. *Environmental Performance, Regulations and Technologies in the Pulp and Paper Industry 1996*, multi-client study, EKONO Inc./Duoplan Oy, Helsinki, 1996.

APPENDIX 2

Appendix 2.

1	**Environmental Management Systems (EMS)**	**208**
1.1	General	208
1.2	ISO 14000 Standards	210
1.3	Eco-Management and Audit Scheme (EMAS)	212

Appendix 2.

1 Environmental Management Systems (EMS)

1.1 General

Today, industrial activities are increasingly striving to achieve sound environmental performance by controlling the impact of their production or services on the environment. A cornerstone in this task is to draw up main environmental objectives and an environmental policy for each organization. Environmental policy is part of the Environmental Management System (EMS) which is presented in Fig. 1 (based on ISO 14001).

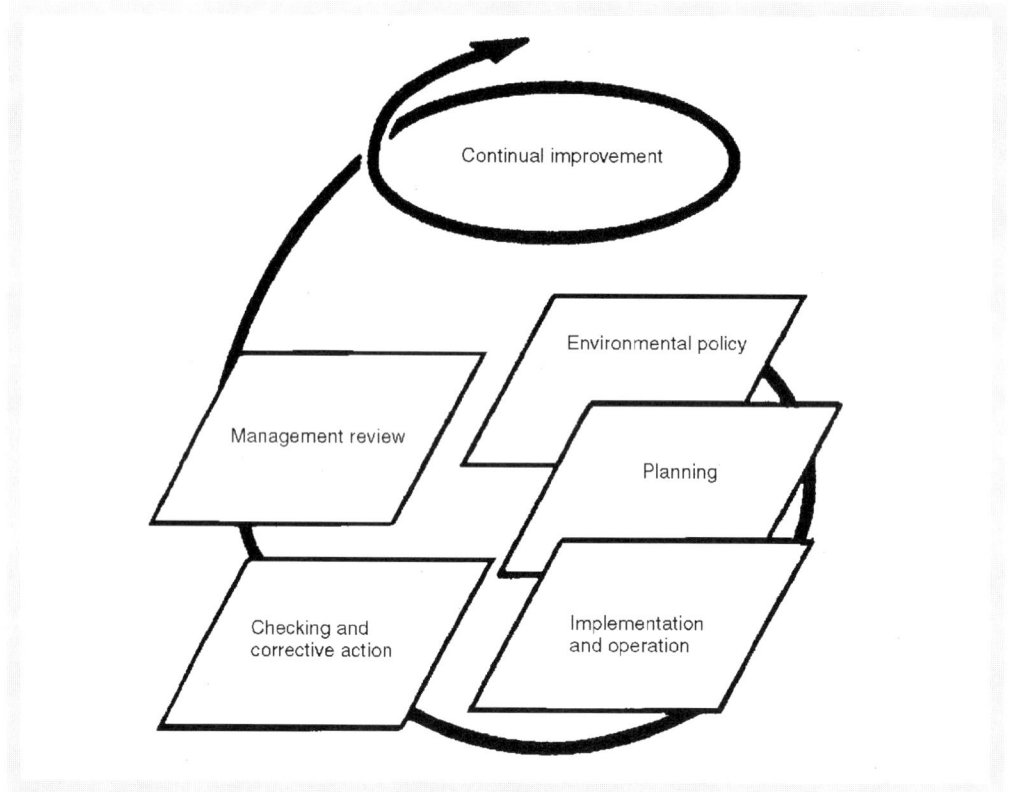

Figure 1. Environmental management system model (ISO 14001).

During the 1990s, International Standards (14000 series by ISO Technical Committee 207) and Regional Schemes (Eco-Management and Audit Scheme, EMAS, by the European Union) were developed for harmonizing the EMS procedure on a voluntary basis. National EMS standards exist also, for example, in the United Kingdom (BS 7750), Ireland, France, and Spain, but after a transition stage these standards will be replaced in most cases with International Standards or EMAS.

The EMS is defined as follows:

- ISO 14001:"that part of the overall management system which includes organizational structure, planning activities, responsibilities, practices, procedures, processes and resources for developing, implementing, achieving, reviewing and maintaining environmental policy"

- EEC/1836/93 (EMAS):"that part of the overall management system which includes the organizational structure, responsibilities, practices, procedures, processes and resources for determining and implementing environmental policy"

Common features of the above procedures are voluntary commitment and continual improvement of the EMS. Major differences between ISO 14001 and EMAS at present are that ISO 14001 can be applied to all kinds of organizations, but EMAS is established for industrial sites only. EMAS also differs from ISO 14001 in that it requires each organization to present a public statement on its environmental performance but ISO 14001 does not.

Some general benefits for a company having an EMS system are listed in ISO 14004, of which the following are crucial:

- Assuring customers of commitment to demonstrable environmental management
- Maintaining good public/community relations
- Satisfying investor criteria and improving access to capital
- Enhancing image and market share
- Meeting vendor certification criteria
- Conserving input materials and energy
- Facilitating the attainment of permits and authorizations
- Improving industry-government relations.

EMS procedures according to ISO 14000 Standards and EMAS are summarized below.

Appendix 2

1.2 ISO 14000 Standards

ISO 14000 standards have been developed and issued by ISO TC 207 in the following areas:

- Standards for organization evaluation:
    - Environmental management systems (EMS)
    - Environmental auditing
    - Environmental performance evaluation
- Standards for product and process evaluation:
    - Environmental labeling
    - Life cycle assessment (LCA)
    - Environmental aspects in product standards
    - Terms and definitions.

The standards for organization evaluation have been adopted and are briefly summarized below. LCA and Environmental Labeling standards are at the draft stage and will be given later in the relevant paragraphs.

The following ISO standards have been approved:

ISO 14001:1996	Environmental management systems – Specification with guidance for use
ISO 14004:1996	Environmental management systems – General guidelines on principles, systems, and supporting techniques
ISO 14010:1996	Guidelines for environmental auditing – General principles
ISO 14011:1996	Guidelines for environmental auditing – Audit procedures – Auditing of environmental management systems
ISO 14012:1996	Guidelines for environmental auditing – Qualification criteria for environmental auditors

ISO 14001

ISO 14001 describes the requirements for certification/registration and/or self-declaration of an organization's environmental management system.

According to the scope of the standard, "this standard specifies requirements for an environmental management system, to enable an organization to formulate a policy and objectives taking into account legislative requirements and information about significant environmental impacts..."

Fig. 1 illustrates the EMS model based on ISO 14001.

Table 1 shows the contents of ISO 14001.

Table 1. ISO 14001 (Environmental management systems. Specification with guidance for use).

1	Scope
2	Normative references
3	Definitions
4	Environmental management system requirements
4.1	General requirements
4.2	Environmental policy
4.3	Planning
4.3.1	Environmental aspects
4.3.2	Legal and other requirements
4.3.3	Objective and targets
4.3.4	Environmental management program(s)
4.4	Implementation and operation
4.4.1	Structure and responsibility
4.4.2	Training, awareness, and competence
4.4.3	Communication
4.4.4	Environmental management system documentation
4.4.5	Document control
4.4.6	Operational control
4.4.7	Emergency preparedness and response
4.5	Checking and corrective action
4.5.1	Monitoring and measurement
4.5.2	Nonconformance and corrective and preventive action
4.5.3	Records
4.5.4	Environmental mangement system audit
4.6	Management review
ANNEX A	(Informative) Guidance on the use of the specification
ANNEX B	(Informative) Links between ISO 14001 and ISO 9001
ANNEX C	Bibliography

ISO 14004

ISO 14004 includes examples, descriptions, and options that aid both in the implementation of an EMS and in strengthening its relation to the overall management of the organization.

Appendix 2

ISO 14004 describes in detail the following key principles of EMS:

- Principle 1: Commitment and policy
- Principle 2: Planning
- Principle 3: Implementation
- Principle 4: Measurement and evaluation
- Principle 5: Review and improvement.

The guidelines of ISO 14004 are intended for use as a voluntary, internal management tool and are not intended to be used as EMS certification/registration criteria.

ISO 14010

ISO 14010 is intended to guide organizations, auditors, and their clients on the general principles common to the conduct of environmental audits. It contains requirements for an environmental audit, general principles including audit criteria and reliability, and an audit report.

ISO 14011

ISO 14011 establishes audit procedures that provide for the planning and conduct of an audit of an EMS to determine conformance with EMS audit criteria.

ISO 14012

ISO 14012 provides guidance on qualification criteria for environmental auditors and lead auditors and is applicable to both internal and external auditors.

1.3 Eco-Management and Audit Scheme (EMAS)

On June 29, 1993, the Council of the European Communities adopted the regulation for a voluntary Eco-Management and Audit Scheme applicable to companies in the industrial sector (EEC 1836/93). At present, EMAS has been adopted in all EU member countries.

The objective of the scheme is presented in Article 1:

a) The establishment and implementation of environmental policies, programs, and management systems by companies, in relation to their sites

b) The systematic, objective, and periodic evaluation of the performance of such elements

c) The provision of information on environmental performance to the public.

The key Articles of the EMAS regulation are:

- Article 4:   Auditing and validation
- Article 5:   Environmental statement
- Article 6:   Accreditation and supervision of environmental verifiers
- Article 7:   List of accredited environmental verifiers
- Article 8:   Registration of sites
- Article 10:   Statement of participation
- Article 12:   Relationship with national, European and international standards
- Article 18:   Competent bodies.

Figure 2. The EMAS procedure.

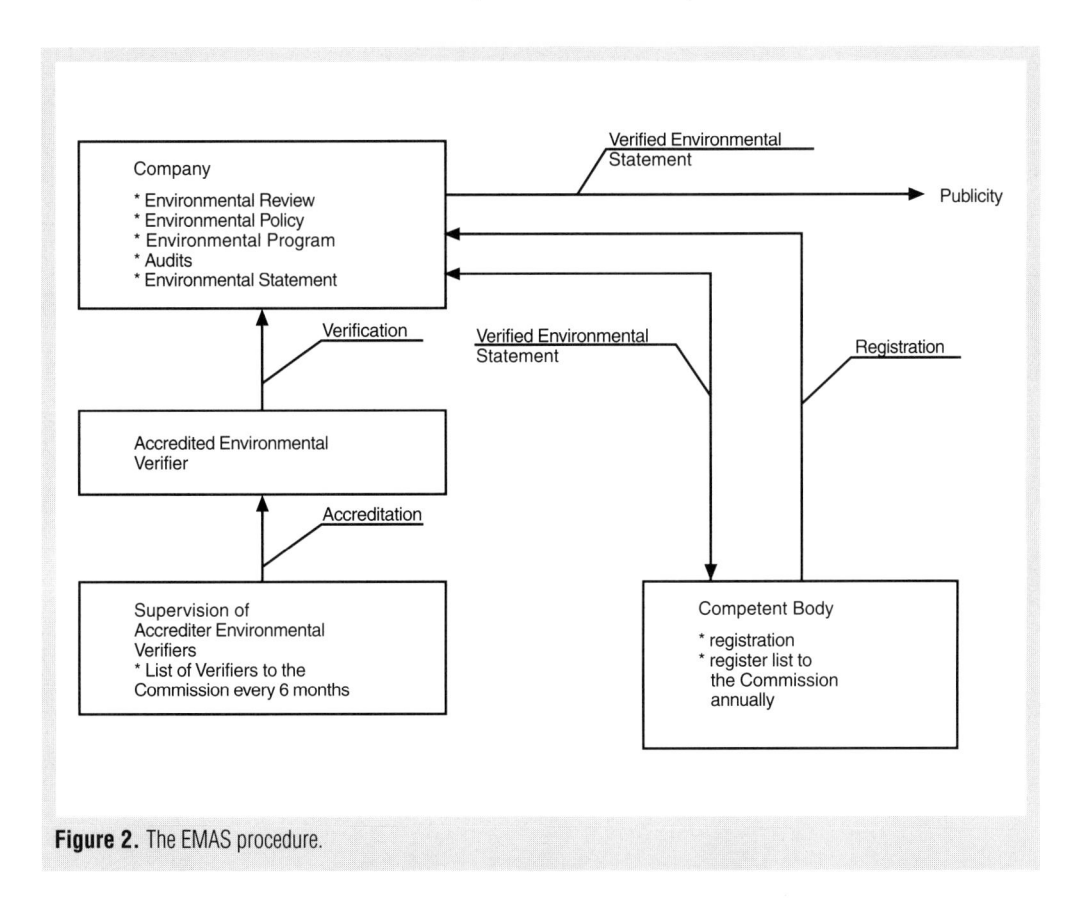

Fig. 2 shows the EMAS procedure.

Appendix 2

The procedure which an individual company is obliged to carry out for EMAS registration approval can be summarized step-by-step as follows:

1) The company must carry out **a preliminary environmental review**. This is the basic evaluation on the level of environmental measures and of the impact on the environment.

2) The company must have its own **environmental policy**, to which the management is committed.

3) Based on Item Nos. 1 and 2, the company shall prepare **an environmental program** which includes specified quantitative goals and the schedule of the program .

4) The company must have an **environmental management system** to show organizational responsibilities in environmental management, etc.

5) The company must prepare an **environmental statement** based on the preliminary environmental review and later after each **audit cycle**. The environmental statement will be verified by an **accredited environmental** verifier and after that the statement will be released publicly.

6) After the above-mentioned stages, the company is able to apply for **registration** by a **competent body**.

7) After approval for EMAS registration, the company must update its environmental policy and environmental program by regular inspections or **audits** at intervals no longer than three years.

APPENDIX 3

Appendix 3.

1	**Life Cycle Assessment (LCA)**	**216**
1.1	General	216
1.2	ISO/DIS 14040 (Environmental Management – Life Cycle Assessment – Principles and Framework)	217

Appendix 3.

1 Life Cycle Assessment (LCA)

1.1 General

As the interest and public awareness of the impact of industrial activities on the environment increased during the 1970s and 1980s, the need for a system or tool to evaluate and compare various production options with respect to their environmental impact became evident. One of the techniques being developed for this purpose is Life Cycle Assessment (LCA).

LCA is a technique for assessing the environmental aspects and potential impact associated with a product by:

- Compiling an inventory of the relevant inputs and outputs of a system
- Evaluating the potential environmental impact associated with those inputs and outputs
- Interpreting the results of the inventory and impact phases in relation to the objectives of the study.

Fig. 1 shows the main phases of an LCA (according to ISO/DIS 14040).

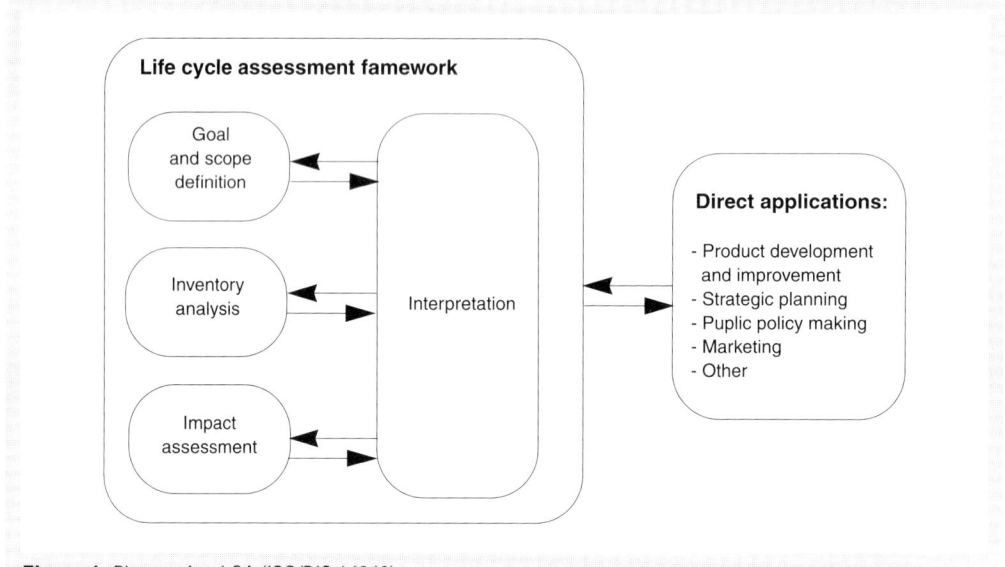

Figure 1. Phases of an LCA (ISO/DIS 14040).

Currently, industrial enterprises use LCA to seek ways of reducing the environmental impact of their products and processes. LCA can be used as a tool in the following applications:

- Assisting in decision-making in industry, e.g., for strategic planning, product or process design, or redesign
- Assisting in decision-making in government (e.g., strategic) planning leading to regulations or research and development funding
- Assisting in decision-making in nongovernmental organizations for strategic planning, priority setting, etc.
- Selection of relevant environmental performance indicators
- Marketing (e.g., for an environmental claim or eco-labeling scheme).

LCA is one of various environmental management techniques (e.g., risk assessment, environmental performance evaluation, site-related environmental auditing, and environmental impact assessment) and may not be the best suitable technique to be used in all cases. LCA typically does not address the economic or social aspects of a process or a product.

At present, there are a number of different models and computer programs for LCA in use, e.g., for the energy and packaging industries. The models are still difficult to compare as the scope, boundaries, and level of detail can vary considerably depending on the subject and use of the study. As the methodologies of LCA are still evolving and their general usefulness still under development, only ISO Draft Standard for LCA will be discussed in this context.

1.2 ISO/DIS 14040 (Environmental Management – Life Cycle Assessment – Principles and Framework)

The ISO/DIS 14040 draft standard (voting terminated on Nov. 13, 1996) provides principles, a framework, and some methodological requirements for conducting LCA studies. Complementary standards, which are in the draft stage, for the various phases of LCA, are:

- ISO 14041:   Environmental management – Life cycle assessment – Goal and scope definition and life cycle inventory analysis
- ISO 14042:   Environmental management – Life cycle assessment – Life cycle impact assessment
- ISO 14043:   Environmental management – Life cycle assessment – Life cycle interpretation.

According to ISO/DIS 14040, the key features of the LCA methodology are:

- LCA studies should systematically and adequately address the environmental aspects of product systems, from raw material acquisition to final disposal.

Appendix 3

- The depth of detail and time frame of an LCA study may vary to a large extent, depending on the definition of goal and scope.
- The scope, assumptions, data quality parameters, methodologies, and output of LCA studies should be transparent and understandable. LCAs should discuss and document the data sources and be clearly and appropriately communicated.
- Provisions should be made, depending on the intended application of the LCA study, to respect confidentiality and proprietary matters.
- LCA methodology should be amenable to the inclusion of new scientific findings and improvements in the state-of-the-art of the methodology.
- Specific requirements are applied to LCA studies which are used to make a comparative assertion that is disclosed to the public.
- There is no scientific basis for reducing LCA results to a single overall score or number, since trade-offs and complexities exist for the systems analyzed at different stages of their life cycle.
- There is no single method for conducting LCAs. Organizations should have flexibility to implement LCA practically as established in ISO/DIS 14040, based on the specific application and the requirements of the user.

The framework for LCA is shown in Fig. 1. Some of the main concepts of the LCA phases are presented below:

- Goal and scope definition

The goal and scope must be clearly defined, e.g., the purpose, system description, boundaries, data requirements, assumption, and limitations

- Life cycle inventory analysis

Inventory analysis involves data collection and calculation procedures to quantify the relevant inputs and outputs of a product system. These inputs and outputs may include the use of resources and releases to air, water, and land associated with the system. Interpretations may be drawn from these data, depending on the goals and scope of the LCA. These data also constitute the input for life cycle impact assessment.

The process of conducting an inventory analysis is iterative. Sometimes issues may be identified that require revisions to the goal or scope of the study.

- Life cycle impact assessment

The impact assessment phase is aimed at evaluating the significance of potential environmental impact using the results of the life cycle inventory analysis. This assessment may include the iterative process of reviewing the goal and scope of the study to determine when the objectives of the study have been met, or to modify the goal and scope if the assessment indicates that they cannot be achieved.

The impact assessment phase may also include:

- Assigning of inventory data to impact categories (classification)
- Modeling of the inventory data within impact categories (characterization)
- Possibly aggregating the results in very specific cases when meaningful (valuation).

There is subjectivity in the life cycle impact assessment phase such as the choice, modeling, and evaluation of impact categories. Therefore, transparency is critical to impact assessment to ensure that assumptions are clearly described and reported.

- Life cycle interpretation

Interpretation is the phase of LCA in which the findings from the inventory analysis and the impact assessment are <u>combined</u> together.

The findings of this interpretation may take the form of conclusions and recommendations to decision-makers, consistent with the goal and scope of the study. The interpretation phase may involve the iterative process of reviewing and revising the scope of the LCA, as well as the nature and quality of the data collected in line with the defined goal.

The findings of the interpretation phase should reflect the results of any sensitivity and uncertainty analysis that is performed.

APPENDIX 4

Appendix 4.

1	**Environmental Labeling (Eco-labeling)**	**221**
1.1	General	221
1.2	Multi-criteria labels	222
1.3	Single-Criterion Labels	224
1.4	Eco-Profiles	224
1.5	Self-declaration environmental claims and symbols	224

Appendix 4.

1 Environmental Labeling (Eco-labeling)

1.1 General

Market forces and consumers currrently require more information about the environmental aspects and impact of industrial products and services. This stress has given impetus to the development of eco-label schemes throughout the world. The use of eco-labels is usually voluntary.

The main objectives of eco-labeling schemes are:

- To promote such development of industrial products that takes environmental viewpoints into consideration in addition to economic and quality matters
- To guide consumers to favor products that cause the least pollution to the environment.

The concept of environmental labeling includes a variety of categories, such as:

- Multi-criteria labels
- Single-criterion labels
- Eco-profiles
- Producers' self-declaration claims
- Particular symbols.

Multi-criteria and single-criterion labels and eco-profiles must usually be certified by a third party, i.e., some independent body that checks and evaluates the environmental impact of a product to be labeled. After evaluation, the third party will decide whether the product meets the criteria for using the label.

Self-declaration claims and symbols can be used by a company by simply adopting them, and no third party is needed for certification.

In the pulp and paper industry, eco-label schemes have been developed both on a national and an international basis. Eco-labels are currently used in North America, Europe, Scandinavia, and also by individual countries. In order to harmonize environmental labeling procedures, ISO has started preparatory work on producing standards on general principles for environmental labeling (ISO/CD 14020 and 14024) and self-declaration environmental claims (ISO/CD 14021 and 14022).

Appendix 4

Other international organizations (e.g., UNEP, UNCTAD, OECD, and GATT) have a significant role in developing global eco-label systems which can acquire international relevance and acceptance.

The main types of eco-labels and some major eco-label schemes are briefly discussed below.

1.2 Multi-criteria labels

Multi-criteria labeling is typically voluntary and certification by a third party is necessary.

The criteria on which the present eco-labels are based include pollution loads on the environment (e.g., COD, AOX, phosphorus, sulfur, and nitrogen emissions), energy requirements (energy used in production and purchased energy), and other specific requirements (e.g., share of recycled fiber in product).

Examples of major multi-criteria labels are:

- The Nordic Swan

In 1989, the Nordic Council of Ministers established the Nordic environmental labeling scheme. Currently, criteria for printing and tissue paper grades exist. The scoring system in both categories is based upon COD, phosphorus, sulfur, and nitrogen oxide emissions (for printing papers, also AOX). Approval of tissue paper for the Nordic Swan label requires a total point score of 4 or less. For printing papers, the total score must be less than or equal to $4 + K$ (K = amount of chemical pulp, as $t_{90\%}/t_{90\%}$ paper).

Products are awarded the label in one country and are allowed to use the label in the other participating countries upon payment of a fee.

- European Community Eco-labels

With regard to the pulp and paper industry, the following eco-labels are applied at present:

- Eco-label for toilet paper (1994)
- Eco-label for kitchen rolls (1994)
- Eco-label for copying paper (1996).

In the case of toilet paper and kitchen rolls, criteria for eco-label evaluation is based on:

- Consumption of renewable resources
- Consumption of nonrenewable resources
- Emission of carbon dioxide
- Emission of sulfur/sulfur dioxide
- Emission of organics to water (COD)
- Emission of chlorinated organics to water (AOX)
- Emission of waste.

To obtain the label for toilet paper, total points according to a specific calculation equation must not exceed 7.5, and for kitchen rolls, 6.5.

Criteria for the copying paper eco-label are:

- Reduction of water pollution: COD < 30 kg/a.d. metric ton, AOX < 0.30 kg/a.d. metric ton.
- Reduction of sulfur emissions: total sulfur emissions from the production both pulp and paper shall not exceed 1.5 kg S/a.d. metric ton.
- Saving of energy: total energy consumption for the whole of pulp and paper production shall not exceed 30 G Joule/a.d. metric ton of pulp and paper. Purchased energy shall not exceed 18 G Joule/a.d. metric ton of pulp and paper.
- Commitment to safeguard forests and good environmental management practices (e.g., EMAS or ISO 14001).
- Blue Angel (Germany)

The scheme is administered by the German Quality Control Institute and covers over 60 product groups, of which recycled paper is one. The standard for recycled products requires 100% recycled fiber as raw material.

- Stichting Milieukeur (the Netherlands)

In 1993 the Dutch Eco-Labeling Organization set criteria for writing paper and notepaper. A fee is charged on application. The criteria for the Dutch eco-label are similar to those of the German Blue Angel. The standards for writing and note papers do not allow for any virgin fiber content.

- Bra Miljöval (Sweden)

In 1993 the Swedish Society for Nature Conservation (Naturskyddföreningen) launched new criteria for their environmental label scheme. Requirements for the labeling are:

- Totally chlorine free (TCF) bleaching is obligatory
- Chelating agents, additives, etc., should be easily degradable (according to OECD standard 301)
- COD and sulfur emission limit values for various pulping processes.
- Environmental Choice (Canada)

In 1990, the Environmental Choice Program (ECP) set standards for newsprint and fine paper, etc. To qualify for the Environmental Choice label, newsprint must contain a minimum of 40% recovered paper, of which 25% must be old newspapers. Fine papers must include at least 50% recovered paper, including a minimum of 10% post-consumer recovered paper.

Appendix 4

In 1995, the ECP was reorganized and new criteria were set for tissue and towel paper grades resembling the calculation system in the equivalent EU eco-label. There are five parameters (each scored separately) in the ECP labeling system:

- Consumption of resources (fibers + nonfibrous additives)
- Consumption of energy in pulping and paper making (points ranged from 0 for < 24 GJ/metric ton of paper, to 8 for > 52 GJ/metric ton of paper)
- COD of the wastewater discharge (range of points awarded from 0 for < 5 kg/metric ton of paper to 8 for > 60 kg/metric ton of paper)
- Sub-lethal toxicity emission factor (TEF) (range of points from 0 for TEF < 50 m^3/metric ton of paper to 8 for TEF > 200 m^3/metric ton of paper)
- Net solid waste production.

In order to qualify for an eco-label, products must score no more than 4 points. The manufacturer must also attest a proactive program to minimize packaging waste.

1.3 Single-Criterion Labels

Single-criterion labels often use only one stage of the life-cycle of the product as a criterion. A single criterion can still be divided into various detailed sub-criteria. A typical example is the certification of sustainable forest management. In this label scheme, several sub-criteria are used, but labeling is concentrated mainly on wood extraction and does not include end-products (paper, furniture, etc.).

The Forest Stewardship Council (FSC) has developed principles and criteria for forest management, as well as guidelines for accreditation and certification. The FSC is in the process of accrediting the first certifiers in the United State and in the United Kingdom.

Regional certification and labeling schemes on sustainable forest management are also being developed, for example, in the EU.

1.4 Eco-Profiles

Eco-profiles are used to gather individual claims into a set of claims representing the various environmental aspects of a product's life-cycle. Eco-profiles differ from multi-criteria labels in that they are intended to give information on products' environmental impact without valuing them. Thus the eco-profiles are not exclusive.

1.5 Self-declaration environmental claims and symbols

Examples of self-declaration labeling are products which are recyclable, refillable, compostable, etc. These claims can be in the form of a symbol, but they can also be presented as single statements.

ISO is at present preparing standards and verification methodologies for self-declaration procedure. These standards are at the draft stage (ISO/CD 14021 and 14022).

APPENDIX 5

Appendix 5.

1	**Analysis methods**	**226**
1.1	Wastewater	226
1.2	Freshwater	226
1.3	Air quality	227
1.4	Air quality. Stationary source emission	227
1.5	Waste management	227

Appendix 5.

1 Analysis methods

1.1 Wastewater

Wastewater parameter	Finnish standard method	Other corresponding standard methods	
	SFS	SCAN	ISO
Total dissolved solids and ash content	3 008		
Suspended solids	3 037		
Color			7 887
Sodium and potassium	3 017		
Calcium and magnesium	3 018		
Iron	3 047		
COD	5 504		
BOD	5 508		5 815
Phosphorus	3 026 (modified)		
Nitrogen (Kjeldahl)	5 505		
Ammonium			5 664
TOC	8 245 (SFS-ISO)		
AOX		9:89	
Toxicity tests			
- with algae	5 072		
- with rainbow trout	5 073		

1.2 Freshwater

Water parameter	Finnish standard method	Other corresponding standard methods
	SFS	ISO
Water quality-Guide to analytical quality control for water analysis		ISO/TR 13 530
Permanganate index		8 467
Alkalinity in water		9 963-(1-2)
Coliform bacteria	4 089	
Nitrite nitrogen	3 029	
The sum of nitrite and nitrate	3 030	13 395
Manganese	3 048	
Aluminium	5 502	
Arsenic		11 969
Council Directive of 15 July 1980 relating to the quality of water intented for human consumption: 80/778/EWG*80/778/EEC*80/778/CEE		

1.3 Air quality

Air quality parameter	Finnish standard method SFS	Other corresponding standard methods ISO
Bioindication, Moss bag method	5 794	
Organic vapours in workplace air	3 861	
Suspended particulates	3 863	
Sulphur dioxide	3 864 and 5 265	
Hydrogen sulfide	5 293	
Nitrogen oxides	5 425	7 996

1.4 Air quality. Stationary source emission

Water parameter	Finnish standard method SFS	Other corresponding standard methods EN	ISO
Sulfur dioxide	5 265		7 934 and 11 632
Particulate emissions	3 866		9 096
Nitrogen oxides			10 849 and 11 564
PCDDs/PCDFs		1 948-(1-3)	
HCl		1 911-(1-3)	

1.5 Waste management

	Finnish standard method SFS	Other corresponding standard methods EN
Mobile waste containers		840-(1-6)

Conversion factors

To convert numerical values found in this book in the RECOMMENDED FORM, divide by the indicated number to obtain the values in CUSTOMARY UNITS. This table is an excerpt from TIS 0800-01 "Units of measurement and conversion factors." The complete document containing additional conversion factors and references to appropriate TAPPI Test Methods is available at no charge from TAPPI, Technology Park/Atlanta, P. O. Box 105113, Atlanta GA 30348-5113 (Telephone: +1 770 209-7303, 1-800-332-8686 in the United States, or 1-800-446-9431 in Canada).

Property	To convert values expressed in RECOMMENDED FORM	Divide by	To obtain values expressed In CUSTOMARY UNITS
Area	square centimeters [cm^2]	6.4516	square inches [in^2]
	square meters [m^2]	0.0929030	square feet [ft^2]
	square meters [m^2]	0.8361274	square yards [yd^2]
Energy	joules [J]	1.35582	foot pounds-force [ft • lbf]
	joules [J]	9.80665	meter kilograms-force [m • kgf]
	millijoules [mJ]	0.0980665	centimeter grams-force [cm • gf]
	kilojoules [kJ]	1.05506	British thermal units, Int. [Btu]
	megajoules [MJ]	2.68452	horsepower hours [hp • h]
	megajoules [MJ]	3.600	kilowatt hours [kW • h or kWh]
	kilojoules [kJ]	4.1868	kilocalories, Int. Table [kcal]
	joules [J]	1	meter newtons [m • N]
Length	nanometers [nm]	0.1	angstroms [Å]
	micrometers [μm]	1	microns
	millimeters [mm]	0.0254	mils [mil or 0.001 in]
	millimeters [mm]	25.4	inches [in]
	meters [m]	0.3048	feet [ft]
	kilometers [km]	1.609	miles [mi]
Mass	grams [g]	28.3495	ounces [oz]
	kilograms [kg]	0.453592	pounds [lb]
	metric tons (tonne) [t] (= 1000 kg)	0.907185	tons (= 2000 lb)
Mass per unit volume	grams per liter [g/L]	7.48915	ounces per gallon [oz/gal]
	kilograms per liter [kg/L]	0.119826	pounds per gallon [lb/gal]
	kilograms per cubic meter [kg/m^3]	1	grams per liter [g/L]
	megagrams per cubic meter [Mg/m^3]	27.6799	pounds per cubic inch [lb/in^3]
	kilograms per cubic meter	16.0184	pounds per cubic foot [lb/ft^3]
Power	watts [W]	1.35582	foot pounds-force per second [ft • lbf/s]
	watts [W]	745.700	horsepower [hp] = 550 foot pounds-force per second

Conversion factors

Property	To convert values expressed in RECOMMENDED FORM	Divide by	To obtain values expressed In CUSTOMARY UNITS
	kilowatts [kW]	0.74570	horsepower [hp]
	watts [W]	735.499	metric horsepower
Pressure, stress, force per unit area	kilopascals [kPa]	6.89477	pounds-force per square inch [lbf/in^2 or psi]
	Pascals [Pa]	47.8803	pounds-force per square foot [lbf/ft^2]
	kilopascals [kPa]	2.98898	feet of water (39.2°F) [ft H$_2$O]
	kilopascals [kPa]	0.24884	inches of water (60°F) [in H$_2$O]
	kilopascals [kPa]	3.38638	inches of mercury (32°F) [in Hg]
	kilopascals [kPa]	3.37685	inches of mercury (60°F) [in Hg]
	kilopascals [kPa]	0.133322	millimeters of mercury (0°C) [mm Hg]
	megapascals [Mpa]	0.101325	atmospheres [atm]
	Pascals [Pa]	98.0665	grams-force per square centimeter [gf/cm^2]
	Pascals [Pa]	1	newtons per square meter [N/m^2]
	kilopascals [kPa]	100	bars [bar]
Speed	meters per second [m/s]	0.30480	feet per second [ft/s]
	millimeters per second [mm/s]	5.080	feet per minute [ft/min or fpm]
Thickness or caliper	micrometers [μm]	25.4	mils [mil] (or points or thousandths of an inch)
	millimeters [mm]	0.0254	mils [mil] (or 0.001 in.)
	millimeters [mm]	25.4	inches [in]
Viscosity, kinematic	centistokes [cSt]	1	square millimeters per second [mm^2/s]
Volume flow rate	liters per minute [L/min]	3.78541	gallons per minute [gal/min]
	liters per second [L/s]	28.31685	cubic feet per second [ft^3/s]
	cubic meters per second [m^3/s]	0.0283169	cubic feet per second [ft^3/s]
	cubic meters per hour [m^3/h]	1.69901	cubic feet per minute [ft^3/min or cfm]
	cubic meters per second [m^3/s]	0.76455	cubic yards per second [yd^3/s]
	cubic meters per day [m^3/d]	0.00378541	gallons per day [gal/d]
Volume, solid	cubic centimeters [cm^3]	16.38706	cubic inches [in^3]
	cubic meters [m^3]	0.0283169	cubic feet [ft^3]
	cubic meters [m^3]	0.764555	cubic yards [yd^3]
	cubic millimeters [mm^3]	1	microliters [μL]
	cubic centimeters [cm^3]	1	milliliters [mL]
	cubic decimeters [dm^3]	1	liters [L]
	cubic meters [m^3]	0.001	liters [L]

Index

A

Abfallgesetz 161, 163
AbwAG 161, 164
AbwasserVwV 162
Act on Environmental Permit Procedures 20
activated carbon 86–87
activated carbon adsorption 86
activated sludge process 68, 70–73, 75–76, 80–81
Air Pollution Control Act 20, 22, 151
Air quality directives 147
alkali metals 172
alkaline extraction of the pulp 35
alkalinity 43, 48
ammonia 89, 101–102, 124
analysis methods 226
AOX 28, 31–33, 37, 65, 151, 155, 159, 165, 179, 196, 201, 222–223
arsenic 194
ash 97, 109, 111, 113–114, 125, 127–128
atmospheric emissions 159, 163, 168, 170, 173, 175–176, 182, 197
audits 138, 140, 212, 214
auxiliary chemicals 31
azides 171

B

BAV reactor 121
BCT 179
belt filter press 109, 114, 116–118, 127
BEP 198–199
best available techniques 18, 150, 198
Best Available Technology 177, 179
biodegradability 78
biological gas treatment 102
biological oxygen demand 32
biomass 68–71, 73–75, 78, 80–82, 86
biorotor 70, 82, 91
biosludge 71, 73–74, 77–80, 84, 109–110, 112–113, 115, 117–119, 121–122, 129

biosorption 74
Blue Angel 223
BMP 179
board 20, 28, 40, 61, 68, 86, 88, 110, 112, 129, 138, 140, 157–158, 162, 165
bottom sludge load 61
BPT 177–179
breakdown of organic matter 68, 77
British Standards 140

C

calcium sulfite pulp 11
Canada 195
capillary water 113–114
carbon monoxide 147
catalytic oxidation 106, 120
centrifuges 116–117
CEPA 196
CERCLA 177–178
chelating agents 200–202, 223
chlorides 44
chlorine chemicals 35
circulation number 71
clarification 44, 47, 50–51, 58–59, 63–64, 67, 82, 85–86, 91, 114
clarifier performance 61
cloro-organics 31
Cluster Rules 178
coagulating agents 44
coated paper 64
CODCr 31–32, 151, 155, 159, 201
color 31–32, 43–44, 47, 62, 64–65, 68, 76, 87, 165, 179
Commission 14, 16, 29, 144, 148, 151, 158, 198, 202
complete mixing 70, 74, 77
concentrated malodorous gases 102, 104
conventional strainers 46
Council of Ministers 14, 151, 202, 204, 206, 222
cut-off value 64–65

Index

D
Dano biostabilizer 121
dilute malodorous gases 102, 104–106
dioxin emissions 124
directorates 14
discharge fee 174–175
discrete settling 58
disinfection 52
dispersion water 51, 62–63
dry debarking 38–39, 201–202
dry injection 100
dust separators 99

E
eco-labeling 217, 221, 223
eco-profiles 221, 224
economic calculations 137
economic controls 14
economic instruments 22–23, 153, 157, 160, 164, 175, 206
EDTA 204
effluent loadings 10–11, 31–32, 38–40, 57
effluent parameters 28, 32, 40, 57
electrostatic precipitator 96–98, 124
elemental chlorine 32–33
EMAS 18, 143, 153, 209, 212–214, 223
EMS 208–212
energy generation 95, 100, 111, 138
environmental action programs 17, 144
environmental aspects 210, 216–217, 221, 224
Environmental Choice 223
environmental controls 13
Environmental impact assessment 29, 140, 144, 153, 175, 194, 197, 217
environmental labels 11, 18, 23
Environmental Management Systems
.... 18, 204, 208, 210
Environmental Permit Procedures Act 29
environmental permits 20, 27–29, 165
environmental policy. 15, 18, 21, 144, 157–158, 160, 176–177, 198, 206, 208, 214
environmental program 214
environmental statement 214
environmental systems 139–140
enzymes 68
European Community Eco-labels 222
European Council 14
European Union 14–15, 17, 143–144, 149, 209
evaporation 32, 35, 38, 87, 89–90, 114, 129, 202
excess sludge 71, 73, 78–79, 82, 127
Executive Authorities 143, 154, 158, 161, 165, 169, 175, 177, 195, 198
extended aeration 73
extended cooking 35, 179

F
fatty acids 37, 57, 76, 80
Federal Water Act 161–162
fiber analysis 114
fiber length 114
fillers 31, 57, 109, 111, 113
filtration 40, 44, 47, 51, 58, 64, 66–68, 82, 85, 114, 116
fine paper 64, 223
Finland...10, 14, 19, 21, 23, 27–29, 32, 39, 43, 47–49, 64–65, 67–68, 74, 110–111, 117–118, 123–124, 127–129, 133, 137–138, 140, 150–151, 153, 202
Finnish Environment Institute 22, 154
fixed bed processes 70
floc load 71, 73, 80
flocculation 44, 47, 49, 58–59, 61, 64, 67–68, 115
flotation 40, 50–51, 58, 62–64, 66–68, 81, 115
flue gas scrubber 100
fluidized bed boilers 124
fluidized bed combustion 124
fluorine 147
Forest Stewardship Council 224
Framework Convention on Climate Change ......
.... 148, 205
Framework Directive 145, 147, 149
France 22, 165, 198, 209
free carbon dioxide 44
Freeness 114
freezing 39, 90
freezing the effluent 39
Froude number 50

G

gas scrubbers 99
Geneva Convention 148, 204
Germany 137, 161, 163, 223
granular carbon 86
grate combustion 124
green liquor dregs 111
Groundwater Directive 145
groundwood 39, 112

H

hardness 43–44, 48
hazardous waste directive 149
heat treatment 118–119
heat value 114
heavy metals 125–126, 128, 145, 147, 155, 170, 179, 194
HELCOM 151, 199–202
High level of protection 15–16, 143, 149
high-load process 73
hindered settling 58
HMIP 169, 171
horizontal clarification 50
hydrocarbons 145, 147, 165
hydrolysis 47, 91, 118

I

impact assessment 29, 140, 144, 153, 175, 194, 197, 216–219
impurities 43, 53, 100, 102
intercellular water 113–114, 118
interpretation 16, 216, 219
inventory analysis 216, 218–219
investment costs 127, 137
ion exchange 44, 53, 89
iron 43–44, 47, 67, 69, 80, 115
Italy 172–173

K

Kappa number 34–35, 37

L

lagoons 70, 81
landfill gases 127
landfilling 109, 113, 125, 127
latex 64
layered composting 121

legislation 11, 14, 16, 18–21, 23, 28, 143–144, 146, 149, 154–155, 159, 161, 165, 169, 172–173, 176–178, 195–196, 198
Licensing Board 157–158
life cycle analysis 11, 23, 138–139
life cycle analysis 11, 23, 138–139
Lignin removal process 88
lignosulphonates 64
lime 48, 67, 91, 95, 97–98, 104, 106, 109, 111, 116, 124, 157, 185
lime kiln 95, 98, 104, 106, 157
location-related environmental permits 28
low-load process 73

M

Maastricht Treaty 15, 17
magnesium 43
malodorous gases 100–106
manganese 43–44, 47, 80
market instruments 23
metal carbonyls 171
methane formation 91
micro-strainers 46
Mindestandforderungen 162
Ministry of the Environment 21–22, 128, 137–138, 154, 161, 163, 165, 176–177, 198
multi-criteria labels 221–222, 224
multi-fuel boilers 99
multicyclone 96, 99

N

National Environmental Policy Act 177
natural alkalinity 48
Natural Resources 15, 143–144, 155, 158
nature conservation 13, 21, 223
newsprint 112, 203, 223
nitrogen 31–32, 57, 70, 76, 78–80, 91, 95, 101–102, 120–121, 128, 147–148, 152, 156–158, 170, 193, 200–201, 203–204, 222
nitrogen and phosphorus removal 70
nitrogen oxides 95, 101–102, 147, 156, 193, 204
noise abatement 18, 20, 28, 133–134, 158
Noise Abatement Act 20

232

Index

Nordic Council of Ministers 151, 202, 204, 206, 222
Nordic Swan 222
normal-load process 73
Norway 159–160, 202
NO_x 157, 193, 197
NSPS 177, 179, 182, 193–194
nutrients 28, 32, 39, 57, 66, 68–69, 74, 76, 80, 122, 128

O

odor 44, 47, 86, 103–104, 106, 122, 127, 152, 156, 165
OECD 23, 137, 222–223
oils 111, 145, 148–149, 160
operating costs 63, 86, 125, 137
organic carbon 69
OSPAR 198
oxidation 32, 44, 47, 79, 90, 105–106, 118–120, 171
oxygen delignification 34–35, 179, 201
oxygen processes 74
ozone 37, 52, 90, 146–148, 193, 197, 205
ozone layer 146–148, 205

P

Packaging and Waste Packaging Directive
.... 19, 149
packed column scrubbers 101
PARCOM Convention 198
particular symbols 221
particulates 101, 147, 156, 193, 197
peripheral speeds 49
permeate 64–65
peroxide 37, 39, 80, 162
peroxide bleaching 39
pesticides 155, 172
phosphorus 31–32, 57, 70, 76, 78, 80, 121, 128, 151, 159, 171–172, 222
pigments 111
pitch content 114, 116
plug flow 70, 74–75
Polluter pays principle 16, 144
Portugal 176
post-pressing 118
potassium 32, 69, 80
Precautionary principle 16, 143
preservatives 11, 111

pressure groundwood 39, 112
primary sludge 110, 118, 120, 129
Principle of prevention 16
PSD 193
PSES 179
PSNS 179
Public Health Act 20
pulp bleaching .. 32, 64–65, 76, 80, 87, 90, 180
pulp yield 32, 39

R

rapid mixing 44, 47, 49, 67
recovery boiler 98, 102, 104–106, 156–157, 159
recycled fiber 23, 40, 112, 162, 222–223
recycling 39–40, 59–60, 81–82, 85, 90, 148–149, 168, 173
reduced sulfur compounds 89, 95, 103, 105
Regulations 15, 17, 20, 29, 128, 133, 143–145, 148, 150–151, 153–155, 157, 159–163, 165, 168–179, 182, 193–199, 206, 217
regulatory controls 11, 14
removal of organic matter 75, 91
resin acids 37, 57, 76, 80
retention time 49–51, 58–59, 61, 71–73, 84, 86
reuse 148–149, 193, 201–202
Reynolds number 50
ridge stack composter 121–122

S

sand filtration 40, 47, 66–67
sawmills 40, 100
scrap metal 111
screens 44–45, 115
screw press 118
self-declaration claims 221
semi-dry method 100
settling 44, 58–59, 61–63, 68, 71, 73–74, 76, 78, 80–81, 114, 128, 174
short-circuiting 61–62
sieves 44
Single European Act 15
Single-criterion labels 221, 224
sludge age 71, 77–78, 80
sludge conditioning 114–115, 118
sludge content 71–72

233

sludge index 71
sludge load 61, 71–72, 77–78
sludge return ratio 72
sludge thickening 115
sludges 63, 84–85, 109–110, 113–115,
    117–119, 121, 123–125, 127–130
sodium 38, 48, 53, 91, 102, 134
sodium aluminate 48
sodium carbonate 48
soil improvement 109, 122–123, 127–129
solvents 106, 111, 155, 171
sorption 69, 75, 80
Source principle 16, 144
Spain 173, 209
stabilization 70, 74, 85, 115–116, 121
step aeration 74
Stichting Milieukeur 223
stripping 89, 104, 106, 187, 201
sulfur balance 99–100
sulfur dioxide 76, 80, 89, 95, 98–102,
    104–106, 140, 147, 151–152, 157,
    222
sulfur emissions 10, 38, 100, 151, 203, 223
surface chemistry 40, 61, 68
surface load 50–51, 59, 61, 66, 72, 115
suspended solids 28, 31–32, 61–64, 66, 71,
    73, 77, 79–80, 85, 90, 110, 120,
    156, 159
sustainable growth 15
Sweden 27, 89, 155, 158, 202, 223

T

TA Luft 161–162
Tabella 172–173
tapered aeration 74
TCF 180, 223
TCLP toxic 194
tertiary air 102, 105–106
The IPPC Directive 18, 149–150
thermal oxidation 105–106
thermomechanical pulp 39
thickening 58, 114–115, 117, 128
thin-layer fermentation 82
tissue 18, 112, 222, 224
TMP 112, 162
Towards Sustainability 144
toxic substances 68–69, 76, 177, 196
toxicity 31–32, 37, 75–76, 194, 196, 224

Trade Effluents 170
transport 17, 20, 28, 63, 128, 134, 148
Treaty 15, 17, 23, 143
Treaty of Amsterdam 15
Treaty of Rome 15
tubular clarifiers 51
turbidity 43

U

ultrafiltration 40, 64–65
United Kingdom 169, 209, 224
United States 74, 120, 127–129, 137, 177, 195
use of natural resources 15, 158

V

vacuum filter 117
vanes 49
venturi scrubber 96–98, 101
vertical clarification 50
Vienna Convention 148, 205
virgin fiber 223
viscosity 49, 58–59, 114
VOC 193
volumetric loading 71

W

waste incineration 123, 147, 163
Waste Management Act .. 20, 22, 153, 161–163
Waste Tax 154
Water Act 19–20, 22, 151, 161–162, 177
Water Decree 27
Water Industry Act 170
Water Pollution Control 14, 18–19, 21–22,
    35, 92, 145–146, 150–151, 154–155,
    159, 161, 172, 195, 198
Water Quality Objectives 170
Water Resources Act 170
Water Rights Appeal Court 28
water supply system 53
wet method 100
wood procurement 133
wood waste 100, 109, 111

Z

Zeta potential 48, 68, 88, 114
Zimpro process 118–119